T0329592

**Learning Automata and Their Applications
to Intelligent Systems**

Learning Automata and Their Applications to Intelligent Systems

JunQi Zhang
Tongji University, Shanghai, China

MengChu Zhou
New Jersey Institute of Technology, New Jersey, USA

IEEE PRESS
WILEY

Published by John Wiley & Sons, Inc., Hoboken, New Jersey.
Published simultaneously in Canada.

For general information on our other products and services or for technical support, please contact our Customer Care Department within the United States at (800) 762-2974, outside the United States at (317) 572-3993 or fax (317) 572-4002.

Wiley also publishes its books in a variety of electronic formats. Some content that appears in print may not be available in electronic formats. For more information about Wiley products, visit our web site at www.wiley.com.

Library of Congress Cataloging-in-Publication Data Applied for:

Hardback ISBN: 9781394188499

Cover Design: Wiley
Cover Image: © Vertigo3d/Getty Images

Set in 9.5/12.5pt STIXTwoText by Straive, Chennai, India

Contents

About the Authors

JunQi Zhang received the PhD degree in computing science from Fudan University, Shanghai, China, in 2007. He became a post-doctoral research fellow and a lecturer in the Key Laboratory of Machine Perception, Ministry of Education in Computer Science, Peking University, Beijing, China, in 2007. He is currently a full professor in the Department of Computer Science and Technology, Tongji University, Shanghai. His current research interests include intelligent and learning automata, reinforcement machine learning, particle swarm optimization, firework algorithm, big data and high-dimensional index, and multimedia data management. He has authored 20+ IEEE Transactions and 60+ conference papers in the above areas. Prof. Zhang was a recipient of the Outstanding Post-Doctoral Award from Peking University.

MengChu Zhou received his BS degree in control engineering from the Nanjing University of Science and Technology, China, in 1983; MS degree in automatic control from the Beijing Institute of Technology, China, in 1986; and PhD degree in computer and systems engineering from Rensselaer Polytechnic Institute, USA, in 1990. He then joined the New Jersey Institute of Technology in 1990 and has been a distinguished professor since 2013. His interests are in intelligent automation, robotics, Petri nets, and AI. He has over 1100 publications including 14 books, 600+ IEEE Transactions papers, and 31 patents. He is a recipient of Humboldt Research Award for US Senior Scientists from Alexander von Humboldt Foundation; Franklin V. Taylor Memorial Award and Norbert Wiener Award from IEEE Systems, Man, and Cybernetics Society; and Edison Patent Award from the Research & Development Council of New Jersey. He is a fellow of IEEE, IFAC, AAAS, CAA, and NAI.

Preface

Stochastic ranking and selection aim to design statistical procedures that select a candidate with the highest mean performance from a finite set of candidates whose performance is uncertain but may be estimated by a learning process based on interactions with a stochastic environment or by simulations. The number of interactions with environments or simulations of candidates is usually limited and needs to be minimized due to limited computational resources. Traditional approaches taken in the literature include frequentist statistics, Bayesian statistics, heuristics, and asymptotic convergence in probability. Novel and recent approaches to stochastic ranking and selection problems are learning automata and ordinal optimization.

Surprisingly, none introduces or studies learning automata and ordinal optimization in the same one book as if these two techniques have no relevance at all. A learning automaton is a powerful tool for reinforcement learning and aims at learning the optimal action that maximizes the probability of being rewarded out of a set of allowable actions by the interaction with a stochastic environment. An update scheme of its state probability vector of actions is critical for learning automata. Its action probability vector plays two roles: (i) deciding when it converges, which is highly related to its used total computing budget, and (ii) allocating computing budget among actions to identify the optimal one, where only ordinal optimization is required. Ordinal optimization has emerged as an efficient technique for simulation optimization. Its underlying philosophy is to obtain good estimate of the optimal action or design while the accuracy of an estimate need not be that high. It is much faster for selecting an outstanding action or design in many practical situations if our goal is to find the best one rather than an accurate performance value of the best one. Therefore, learning automata and ordinal optimization share the common objective and the latter can provide efficient methods as an update scheme of the state probability vector of actions for learning automata.

This book introduces and combines learning automata and ordinal optimization to solve stochastic ranking and selection problems for the first time. This book may serve as a reference for those in the field and as a means for those new to the field for understanding and applying the main reinforcement learning and intelligent optimization approaches to the problems of their interest. This book is both a research monograph for the intended audience including researchers and practitioners, and a main reference book for a first-year graduate course for the graduate students, in the fields of business, engineering, management science, operation management, stochastic control, economics, and computer science. Only basic knowledge of probability and an undergraduate background in the above-mentioned majors are needed for understanding this book's materials.

JunQi Zhang
Shanghai, China

MengChu Zhou
New Jersey, USA and Hangzhou, China

Acknowledgments

From the first author of this book:

I would like to thank all the people who have contributed to this book and the research team at Tongji University for their full dedication and quality research. In particular, I would like to acknowledge the following individuals.

First, I would like to express my great appreciation to this book's co-author, Professor MengChu Zhou, for his inspirational advice and insightful suggestions to help strengthen the visions and concepts of this book.

I would like to thank the significant help from my students Drs. Huan Liu and DuanWei Wu, Mr. Peng Zu, Mr. YeHao Lu, and Mr. YunZhe Wu and for content and material preparations, as well as the research outcomes.

I would like to appreciate the Wiley-IEEE Press for providing the opportunity to publish this book and the esteemed editor and anonymous reviewers for reviewing our work. Special thanks are given to Jayashree Saishankar, Managing Editor at Wiley-IEEE, and Victoria Bradshaw, Senior Editorial Assistant, who kindly and patiently helped us move smoothly during our book writing and preparation period.

I would like to acknowledge the support from Innovation Program of Shanghai Municipal Education Commission (202101070007E00098), Shanghai Industrial Collaborative Science and Technology Innovation Project (2021-cyxt2-kj10), Shanghai Municipal Science and Technology Major Project (2021SHZDZX0100), the Fundamental Research Funds for the Central Universities, the National Natural Science Foundation of China (51775385, 61703279, 62073244, 61876218), and the Shanghai Innovation Action Plan under grant no. 20511100500.

Finally, I truly appreciate the continuous support and endless love from my family, especially from my wife and two children, who always stay with me and make my life colorful and happy.

From the second author of this book:

Numerous collaborations have been behind this book and its related work. It would be impossible to reach this status without the following collaborators, some of whom are already mentioned in the first author's message.

I would like to thank Professors Naiqi Wu (Fellow of IEEE, Macau Institute of Systems Engineering, Macau University of Science and Technology, China), Zhiwu Li (Fellow of IEEE, Macau Institute of Systems Engineering, Macau University of Science and Technology, China), Maria Pia Fanti (Fellow of IEEE, Dipartimento di Elettrotecnica ed Elettronica, Polytechnic of Bari, Italy), Giancarlo Fortino (Fellow of IEEE, Department of Computer Science, Modeling, Electronics and Systems Engineering (DIMES), University of Calabria, Italy), and Keyi Xing (Systems Engineering Institute, Xi'an Jiaotong University, China).

I have enjoyed the great support and love from my family for long. It would be impossible to accomplish this book and many other achievements without their support and love. The work presented in this book was in part supported by FDCT (Fundo para o Desenvolvimento das Ciencias e da Tecnologia) under Grant No. 0047/2021/A1, and Lam Research Corporation through its Unlock Ideas program.

A Guide to Reading this Book

This book is written for senior students, graduate students, researchers, and practitioners in relevant machine learning fields. This book can be a reference book for those studying computer science/engineering, computational intelligence, machine learning, artificial intelligence, systems engineering, industrial engineering, and any field that deals with the optimal design and operation of complex systems with noisy disturbance to apply learning automata to their problem solving. The readers are assumed to have a background of computer programming and discrete mathematics including set theory and predicate logic.

If readers have knowledge of formalization, optimization, and machine learning, it should be easy for them to understand our presented algorithms and problem formalizations. Readers may choose to learn the ideas and concepts of learning automata. Readers may focus on the novel ideas and their related algorithms that are useful for their research and particular applications. If you do not like the ordinal optimization, you may choose to skip it and focus on the learning automata. The following chart should help readers better obtain what they need to read based on what they would learn from this book.

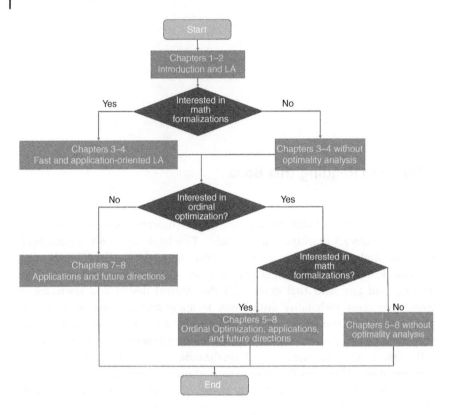

Organization of the Book

To overview the contents, this book first introduces the basic concept of learning automata in Chapter 2. Two pioneering and improved variants of learning automata from the perspectives of convergence and computational cost, respectively, are presented in Chapter 3. Two application-oriented learning automata are given Chapter 4. One discovers and tracks spatiotemporal event patterns. The other solves the problem of stochastic search on a line. Ordinal optimization is another method to solve stochastic ranking and selection problems and is introduced in Chapter 5 through demonstrations of two pioneering variants of optimal computing budget allocation (OCBA). Chapter 6 incorporates learning automata with ordinal optimization to further improve their convergence performance. The role of ordinal optimization is separated from the action probability vector in learning automata. Then, as a pioneering ordinal optimization method, OCBA is introduced into learning automata to allocate the computing budget to actions in a way that maximizes the probability of selecting the true optimal action. The differences and relationships between learning automata and ordinal optimization can be well understood through this chapter. Chapter 7 demonstrates an example to show how both learning automata and OCBA can be used in noisy optimization. Finally, Chapter 8 summarizes the existing applications of learning automata and suggests their future research directions.

1

Introduction

It is an expensive process to rank and select the best one from many complex discrete event dynamic systems that are computationally intensive to simulate. Therefore, learning the optimal system, action, alternative, candidate, or design is a classical problem and has many applications in the areas of intelligent system design, statistics and stochastic simulation, machine learning, and artificial intelligence.

Stochastic ranking and selection of given system designs can be treated as a simulation optimization problem. Its solution requires one to design statistical procedures that can select the one with the highest mean performance from a finite set of systems, alternatives, candidates, or designs. Their mean values are unknown and can be estimated only by statistical sampling, because their reward from environments is not deterministic but stochastic due to unknown noise. It is a classical problem in the areas of statistics and stochastic simulation.

Example 1.1 Finding the most effective drug or treatment from many different alternatives where the economic cost of each sample for testing the effectiveness of the drug is very expensive and risky. Another representative example is the archery competition. How can we select the real champion while using the least number of arrows?

Noise comes mainly from three kinds of uncertainties [1]. (i) Environmental uncertainties: operating temperature, pressure, humidity, changing material properties and drift, etc. are a common kind of uncertainties. (ii) Design parameter uncertainties: the design parameters of a product can only be realized to a certain degree of accuracy because high precision machinery is expensive. (iii) Evaluation uncertainties: the uncertainty happens in the evaluation of the system output and the system performance including measuring errors and all kinds of approximation errors if models instead of the real physical objects are used.

Learning Automata and Their Applications to Intelligent Systems, First Edition.
JunQi Zhang and MengChu Zhou.

1.1 Ranking and Selection in Noisy Optimization

Real-world optimization problems are often subject to uncertainty as variables can be affected by imprecise measurements or just corrupted by other factors such as communication errors. In either case, uncertainty is an inherent characteristic of many such problems and therefore needs to be considered when tailoring meta-heuristics to find good solutions. Noise is a class of uncertainty and corrupts the objective values of solutions at each evaluation [2]. Noise has been shown to significantly deteriorate the performance of different metaheuristics such as Particle Swarm Optimizer (PSO), Genetic Algorithms (GA), Evolutionary Strategies (ES), Differential Evolution (DE), and other metaheuristics.

Example 1.2 Ranking and selection are the critical part in PSO. First, each particle has to compare the fitness of its new position to its previous best and retain the better one. Second, the overall best solution found so far has to be determined as a global best solution to lead the swarm flying and refining the accuracy. Since PSO has a memory that stores the estimated personal best solution of a particle and the estimated global best solution of a swarm, noise leads such memory to be inaccurate over iterations and particles eventually fail to rank and select good solutions from bad ones, which drives the particle swarm toward a wrong direction. Concretely, a noisy fitness function induces noisy fitness evaluations and causes two types of undesirable selection behavior [3]: (i) A superior candidate may be erroneously believed to be inferior, causing it to be eliminated; (ii) An inferior candidate may be erroneously believed to be superior, causing it to survive and reproduce. These behaviors in turn cause the following undesirable effects: (i) The learning rate is reduced. (ii) The system does not retain what it has learnt. (iii) Exploitation is limited. (iv) Fitness does not monotonically improve with generation even with elitism.

Uncertainty has to be taken into account in many real-world optimization problems. For example, in the engineering field, the signal returned from the real world usually includes a significant amount of noise due to the measure error or various uncertainties. It is usually modeled as sampling noise from a Gaussian distribution [4]. Therefore, the noise can be characterized by its standard deviation σ and classified into additive and multiplicative ones. Its impact to function $\breve{F}(x)$ can be expressed as:

$$\hat{F}_{+}(x) = \breve{F}(x) + N(0, \sigma^2), \tag{1.1}$$

$$\hat{F}_{\times}(x) = \breve{F}(x) \times N(1, \sigma^2), \tag{1.2}$$

where $\breve{F}(x)$ represents the real fitness value of solution x, $\hat{F}_{+}(x)$ and $\hat{F}_{\times}(x)$ are illusory fitness values of solution x in additive and multiplicative noisy environments,

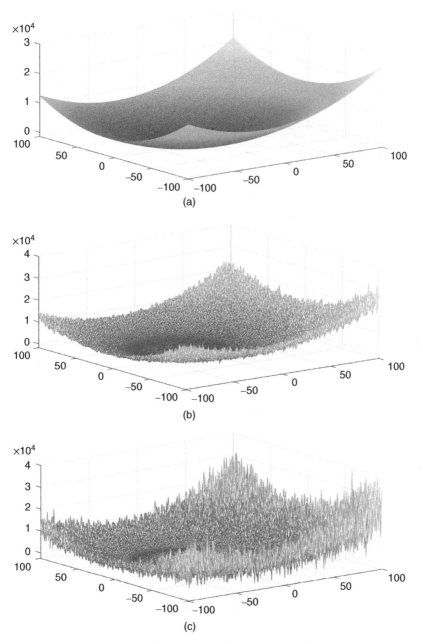

Figure 1.1 3-D map of a Sphere function with additive and multiplicative noise. (a) True. (b) Corrupted by additive noise with $\sigma = 0.1$. (c) Corrupted by multiplicative noise with $\sigma = 0.3$.

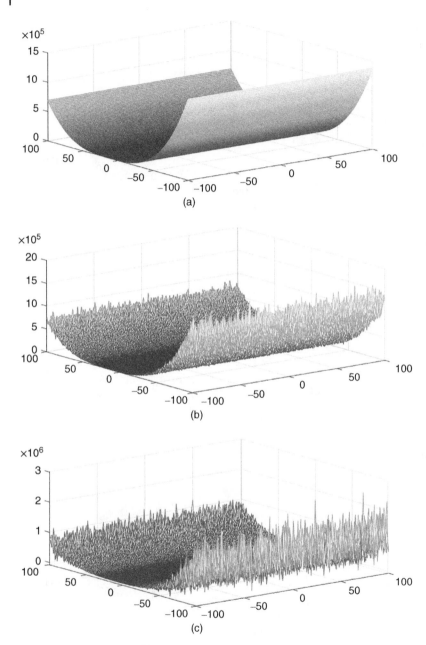

Figure 1.2 3-D map of a Different Powers function with additive and multiplicative noise. (a) True. (b) Corrupted by additive noise with $\sigma = 0.1$. (c) Corrupted by multiplicative noise with $\sigma = 0.3$.

respectively. It is worth noting that they are identical in the environment where $\sigma = 0$. It is obvious that multiplicative noise is much more challenging, since it can bring much larger disturbance to fitness values than additive noise.

Example 1.3 The 3-D maps of functions [5] with additive and multiplicative noise with different σ are illustrated in Figs. 1.1 and 1.2. The figures show that the additional challenge of multiplicative noise over additive noise is a larger corruption of the objective values whose magnitude changes across the search space proportionally to the objective values of the solutions. For multiplicative noise, its impact depends on the optimization objective and the range of objective space. Specifically, on minimization problems whose objective space is only positive, the objective values of better solutions (whose fitness value is small) can be less affected by multiplicative noise. Conversely, in maximization problems, the objective values of better solutions can be more affected [2].

Therefore, if the problem is subject to noise, the quality of the solutions deteriorates significantly. The basic resampling method uses many re-evaluations to estimate the fitness of a candidate solution. Thus, the efficiency of resampling determines the accuracy of ranking and selection of the elite solutions, which are critical to intelligent optimization methods during their evolution to the optimal solutions. Yet the resampling costs too much. Considering a fixed and limited computational budget of function evaluations or environmental interactions, resampling methods better estimate the objective values of the solutions by performing fewer iterations. Intelligent allocation of resampling budget to candidate solutions can save many evaluations and improve the estimation of the solution fitness.

1.2 Learning Automata and Ordinal Optimization

Ranking and selection procedures were developed in the 1950s for statistical selection problems such as choosing the best treatment for a medical condition [6]. The problem of selecting the best among a finite set of alternatives needs an efficient learning scheme given a noisy environment and limited simulation budget, where the best is defined with respect to the highest mean performance, and where the performance is uncertain but may be estimated via simulation. Approaches presented in the literature include frequentist statistics, Bayesian statistics, heuristics [7], and asymptotic convergence in probability [6]. This book focuses on learning automata and ordinal optimization methods to solve stochastic ranking and selection problems.

Investigation of a Learning Automaton (LA) began in the erstwhile Soviet Union with the work of Tsetlin [8, 9] and popularized as LA in a survey paper in 1974 [10]. These early models were referred to as deterministic and stochastic automata operating in random environments. Systems built with LA have been successfully employed in many difficult learning situations and the reinforcement learning represents a development closely related to the work on LA over the years. This has also led to the concept of LA being generalized in a number of directions in order to handle various learning problems.

An LA represents an important tool in the area of reinforcement learning and aims at learning the optimal action that maximizes the probability of being rewarded out of a set of allowable actions by the interaction with a random environment. During a cycle, an automaton chooses an action and then receives a stochastic response that can be either a reward or penalty from the environment. The action probability vector of choosing the next action is then updated by employing this response. The ability of learning how to choose the optimal action endows LA with high adaptability to the environment. Various LAs and their applications have been reviewed in survey papers [10], [11] and books [12–15].

Ordinal Optimization (OO) is introduced by Ho et al. [16]. There are two basic ideas behind it: (i) Estimating the order among solutions is much easier than estimating the absolute objective values of each solution. (ii) Softening the optimization goal and accepting good enough solutions leads to an exponential reduction in computational burden. The Optimal Computing Budget Allocation (OCBA) [17–22] is a famous approach that uses an average-case analysis rather than a worst-case bound of the indifference zone. It attempts to sequentially maximize the probability that the best alternative can be correctly identified after the next stage of sampling. Its procedure tends to require much less sampling effort to achieve the same or better empirical performance for correct selection than the procedures that are statistically more conservative.

Example 1.4 This example shows the objective of ordinal optimization. A system works with one of 10 components that have their own time-to-failure. The objective is to decide which component is the worst one to minimize steady-state system unavailability given budget constraints. We have the budget to perform only 50 component tests to obtain if the tested component leads to a failure. How do we allocate these budget to these 10 components so as to identify the worst one is the objective of ordinal optimization.

The next example shows the reason of softening the optimization goal, i.e., reducing computational burden.

Example 1.5 This example shows the reason of softening the optimization goal and accepting good enough solutions leads to an exponential reduction in computational burden. A class consists of 50 students. The objective is to find the tallest one. However, we do not have to measure all student's accurate height. We can use some sorting algorithms like bubble sorting to identify the tallest student. In this way, we do not need all students' accurate heights. Furthermore, the order of the other students except the tallest one is not important and can be inaccurate. An ordinal optimization method can also be used to identify the tallest student.

1.3 Exercises

1 What is the objective of stochastic ranking and selection?

2 What are the difficulties caused by noise in solving optimization problems?

3 What are the advantages and constraints of resampling in noisy optimization?

4 What is the objective of a learning automaton?

5 What are the basic ideas of ordinal optimization?

6 For additive and multiplicative noise, which one is more difficult to handle and why?

7 In Example 1.5., how to find the tallest student without their accurate height data?

References

1 H.-G. Beyer and B. Sendhoff, "Robust optimization–a comprehensive survey," *Computer Methods in Applied Mechanics and Engineering*, vol. 196, no. 33-34, pp. 3190–3218, 2007.

2 J. Rada-Vilela, "Population statistics for particle swarm optimization on problems subject to noise," Ph.D. dissertation, 2014.

3 A. Di Pietro, L. While, and L. Barone, "Applying evolutionary algorithms to problems with noisy, time-consuming fitness functions," in *Proceedings of Congress on Evolutionary Computation*, vol. 2. IEEE, 2004, pp. 1254–1261.

4 Y. Jin and J. Branke, "Evolutionary optimization in uncertain environments-a survey," *IEEE Transactions on Evolutionary Computation*, vol. 9, no. 3, pp. 303–317, 2005.

5 J. Liang, B. Qu, and P. N. Suganthan, "Problem definitions and evaluation criteria for the CEC 2013 special session on real-parameter optimization," Zhengzhou University, Zhengzhou, China and Nanyang Technological University, Singapore, Technical Report, vol. 201212, 2013.

6 M. C. Fu, "Handbook of simulation optimization," Springer, 2015, vol. 216.

7 Y. Jin, H. Wang, and C. Su, "Data-driven evolutionary optimization integrating evolutionary computation," Machine Learning and Data Science, Springer, Cham, Switzerland, 2021.

8 M. Tsetlin, "On the behavior of finite automata in random media," Automation and Remote Control, pp. 1210–1219, 1961.

9 M. L. Tsetlin, "Automaton theory and the modeling of biological systems," New York: Academic, 1973.

10 K. S. Narendra and M. A. L. Thathachar, "Learning automata: A survey," *IEEE Transactions on Systems, Man, and Cybernetics*, vol. 4, pp. 323–334, 1974.

11 M. A. L. Thathachar and P. S. Sastry, "Varieties of learning automata: An overview," *IEEE Transactions on Systems, Man, and Cybernetics*, vol. 32, pp. 711–722, 2002.

12 S. Lakshmivarahan, "Learning algorithms theory and applications," New York: Springer-Verlag, 1981.

13 K. S. Narendra and M. A. L. Thathachar, "Learning automata: An introduction," Englewood Cliffs, NJ: Prentice-Hall, 1989.

14 K. Najim and A. S. Poznyak, "Learning automata: Theory and applications," New York: Pergamon, 1994.

15 A. S. Poznyak and K. Najim, "Learning automata and stochastic optimization," New York: Springer, 1997.

16 Y. C. Ho, R. S. Sreenivas, and P. Vakili, "Ordinal optimization of discrete event dynamic systems," *Discrete Event Dynamic Systems (DEDS)*, vol. 2, no. 2, pp. 61–88, 1992.

17 H.-C. Chen, C.-H. Chen, and E. Yücesan, "Computing efforts allocation for ordinal optimization and discrete event simulation," *IEEE Transactions on Automatic Control*, vol. 45, no. 5, pp. 960–964, 2000.

18 C.-H. Chen, J. Lin, E. Yücesan, and S. E. Chick, "Simulation budget allocation for further enhancing the efficiency of ordinal optimization," *Discrete Event Dynamic Systems: Theory and Applications*, vol. 10, pp. 251–270, 2000.

19 Y. Ho, Q. Zhao, and Q. Jia, "Ordinal optimization: Soft optimization for hard problems," New York: Springer, 2007.

20 J. Zhang, Z. Li, C. Wang, D. Zang, and M. Zhou, "Approximate simulation budget allocation for subset ranking," *IEEE Transactions on Control Systems Technology*, vol. 25, no. 1, pp. 358–365, 2017.

21 J. Zhang, L. Zhang, C. Wang, and M. Zhou, "Approximately optimal computing budget allocation for selection of the best and worst designs," *IEEE Transactions on Automatic Control*, vol. 62, no. 7, pp. 3249–3261, 2017.

22 J. Zhang, C. Wang, D. Zang, and M. Zhou, "Incorporation of optimal computing budget allocation for ordinal optimization into learning automata," *IEEE Transactions on Automation Science and Engineering*, vol. 13, no. 2, pp. 1008–1017, 2016.

2

Learning Automata

In order to simulate a biological learning process and achieve automatic learning of a machine, Tsetlin [1] et al., first proposed a novel mathematical model called Learning Automaton (LA). LA's performance is optimized by constantly interacting with random environments. Consequently, the optimal action can be selected from the set of alternative ones in the current environment. The optimal action is defined as the action with the highest environmental reward probability in the current environment. LA is one of the reinforcement learning algorithms and has the advantages of being simple and easy to implement, possessing fast random optimization and strong anti-noise ability, and complete convergence. During its development and evolution over several decades, the convergence speed and accuracy of the LA algorithm have been greatly improved. In addition, LAs have also been applied to many application fields, such as graph coloring, random shortest path, distributed computing, wireless network spectrum allocation, image processing, and pattern recognition.

2.1 Environment and Automaton

2.1.1 Environment

An LA consists of two parts: an automaton and an environment. The latter shown in Fig. 2.1 is defined mathematically as a triple $\langle A, B, C \rangle$, which can be explained as follows.

1) $A = \{\alpha_1, \alpha_2, ..., \alpha_r\}$ is a set of actions ($r \geq 2$). The action selected at instant t is denoted by $\alpha(t)$.
2) $B = \{\beta_1, \beta_2, ..., \beta_u\}$ is the output set of possible environmental responses. The environmental response at instant t is denoted by $\beta(t)$. To simplify our discussions, let $B = \{\beta_1, \beta_2\} = \{0, 1\}$. "1" and "0" denote the reward and penalty responses, respectively.

Learning Automata and Their Applications to Intelligent Systems, First Edition.
JunQi Zhang and MengChu Zhou.

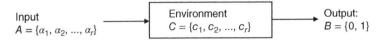

Input
$A = \{\alpha_1, \alpha_2, ..., \alpha_r\}$ ⟶ Environment $C = \{c_1, c_2, ..., c_r\}$ ⟶ Output: $B = \{0, 1\}$

Figure 2.1 An environment.

3) $C = \{c_{ij} = Pr\{\beta(t) = \beta_j | \alpha(t) = \alpha_i\}\}, i \in \mathbb{Z}_r = \{1, 2, ..., r\}, j \in \mathbb{Z}_u = \{1, 2, ..., u\}$
   is the penalty probability matrix that the environment rewards actions.

Since c_{ij} is a conditional probability, the environment is called a random environment. According to the different values of B, the random environment can be divided into three environmental models, i.e., P-model, Q-model, and S-model.

If B is a finite set and its elements are discretely distributed over the unit interval $[0, 1]$, the environmental model is referred to as a Q-model; if B has an infinite number of values, and its elements are continuously distributed in the unit interval $[0, 1]$, the environmental model is referred as an S-model, while an environmental model whose response B contains only output values 0 and 1 is referred as a P-model.

Q- and S-models are more closely related to biological learning and have more practical application values. However, research on a P-model is the basis of studying other more complex environments. Therefore, a P-model is usually more attractive to many scholars.

Considering a P-model, if $c_i = Pr\{\beta(t) = 1 | \alpha(t) = \alpha_i\}$, then we have $1 - c_i = Pr\{\beta(t) = 0 | \alpha(t) = \alpha_i\}$. In this case, the penalty probability can be written as $C = \{c_1, c_2, ..., c_r\}$, where each $c_i, i \in \mathbb{Z}_r$, corresponds to the element α_i of A. If c_{ij} is not a function of time, the environment is known as a stationary environment; otherwise, it is a non-stationary environment.

2.1.2 Automaton

An automaton shown in Fig. 2.2 can be described by a quintuple $\langle A, B, Q, T, G \rangle$.

1) $A = \{\alpha_1, \alpha_2, ..., \alpha_r\}, 2 \leq r < \infty$ is the set of outputs or actions of an automaton.
   The action selected at instant t is denoted by $\alpha(t)$.
2) $B = \{\beta_1, \beta_2..., \beta_u\}$, as an environmental response, is an input set of an automaton. At instant t, it is denoted as $\beta(t)$. B could be infinite or finite.
3) $Q = \{q_1, q_2, ..., q_v\}$ is a state of an automaton. At instant t, the state is denoted by $q(t)$.
4) $T : Q \times B \rightarrow Q$ is a state transfer function of an automaton. T determines how an automaton migrates to a state of $t + 1$ according to the output, input, and the state at an instant t.
5) $G : Q \rightarrow A$ is an output function, which determines how an automaton produces output based on the state.

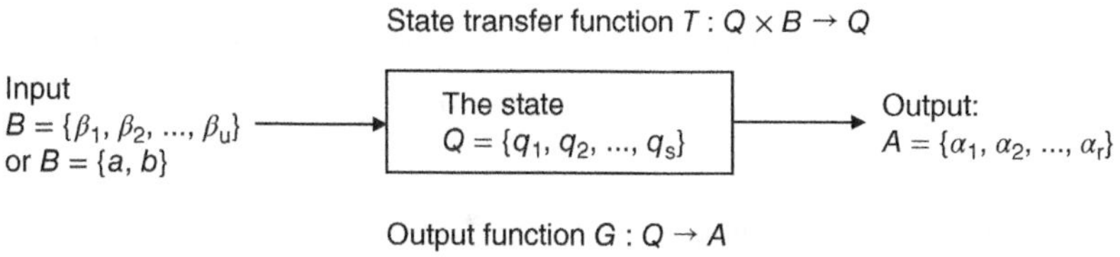

State transfer function $T : Q \times B \rightarrow Q$

Input
$B = \{\beta_1, \beta_2, ..., \beta_u\}$
or $B = \{a, b\}$

The state
$Q = \{q_1, q_2, ..., q_s\}$

Output:
$A = \{\alpha_1, \alpha_2, ..., \alpha_r\}$

Output function $G : Q \rightarrow A$

Figure 2.2 An automaton.

In the above definitions, if A, B, and Q are all finite sets, the automaton is said to be finite.

2.1.3 Deterministic and Stochastic Automata

If the mapping relationship in functions T and G is determined, an automaton is referred to be a deterministic one. More specifically, as a deterministic automaton, given an input and a state, the state and output at the next instant are determined. On the other hand, if any of T or G is random, an automaton is defined as a stochastic one. Running the experiment twice, if T is a random mapping, even if the same state and the same input are given at the same instant, the state at the next instant can be different. Therefore, regardless of whether the output map is deterministic or not, the output of the next instant cannot be determined in advance. Similarly, given the same state, if G is a random mapping, the same action output cannot be guaranteed.

Considering the case of probability, the random mapping of T can be represented as a series of conditional probability matrices:

$$\tau_{il}^k = Pr\{q(t+1) = q_l | q(t) = q_i, \beta(t) = \beta_k\} \, i, l \in \mathbb{Z}_v, \beta_k \in B, \quad (2.1)$$

where t is the transition probability from state q_i to q_l with the environment output β_k at instant t. For each β_k, $\sum_l \tau_{il}^k = 1$. Thereby, we can conclude that T is a Markov matrix, since the state at instant $t + 1$ is only related to the state at instant t.

The random mapping of G can be expressed as a conditional probability matrix:

$$g_{ij} = Pr\{\alpha(t) = \alpha_j | q(t) = q_i\} \, \alpha_j \in A, q_i \in Q, \quad (2.2)$$

where $\sum_j g_{ij} = 1$.

We give two examples to intuitively reflect the difference between deterministic and stochastic automata.

Example 2.1 Consider an automaton with two inputs, i.e., $B = \{0, 1\}$, two outputs, i.e., $A = \{\alpha_1, \alpha_2\}$, and four states, i.e., $Q = \{q_1, q_2, q_3, q_4\}$. A deterministic state transition graph and deterministic output graph are shown in Figs. 2.3 and 2.4, respectively, where each hollow point represents a state, each link

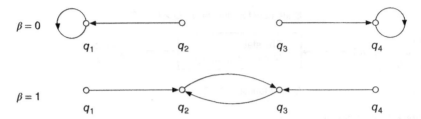

Figure 2.3 The deterministic state transition graph.

Figure 2.4 The deterministic output graph.

represents a state transition, and the arrow on the link indicates the direction of state transition. We can also depict the corresponding deterministic state transition function T and the deterministic output function G with a matrix set and a matrix, respectively. Assume that the entries τ_{ij}^0 and τ_{ij}^1 have definition as follows:

$$\tau_{ij}^\beta = \begin{cases} 1, & \text{if } q_i \to q_j \text{ with an input } \beta = 0 \\ 0, & \text{otherwise} \end{cases}.$$

Then, the deterministic state transition function T can be identified as

$$T(0) = \begin{array}{c} \\ q_1 \\ q_2 \\ q_3 \\ q_4 \end{array} \begin{array}{cccc} q_1 & q_2 & q_3 & q_4 \\ \left[\begin{array}{cccc} 1 & 0 & 0 & 0 \\ 1 & 0 & 0 & 0 \\ 0 & 0 & 0 & 1 \\ 0 & 0 & 0 & 1 \end{array} \right] \end{array} \qquad T(1) = \begin{array}{c} \\ q_1 \\ q_2 \\ q_3 \\ q_4 \end{array} \begin{array}{cccc} q_1 & q_2 & q_3 & q_4 \\ \left[\begin{array}{cccc} 0 & 1 & 0 & 0 \\ 0 & 0 & 1 & 0 \\ 0 & 1 & 0 & 0 \\ 0 & 0 & 1 & 0 \end{array} \right] \end{array}.$$

The sum of each row of the matrices is 1, and the elements in the matrices are either 1 or 0. Therefore, at the present instant, if a state and input are determined, a state at the next instant is determined. Similarly, assume that entry g_{ij} is defined as follows:

$$g_{ij} = \begin{cases} 1, & \text{if } G(q_i) = \alpha_j \\ 0, & \text{otherwise} \end{cases}.$$

The deterministic output function G can be identified as

$$G = \begin{array}{c} \\ q_1 \\ q_2 \\ q_3 \\ q_4 \end{array} \begin{array}{c} \alpha_1 \ \alpha_2 \\ \begin{bmatrix} 1 & 0 \\ 0 & 1 \\ 1 & 0 \\ 0 & 1 \end{bmatrix} \end{array}.$$

Analogously, the sum of all components in each row of the matrix is 1, and the elements in the matrix are either 1 or 0. Therefore, if a state at the present instant is determined, an action at the next instant is determined.

Example 2.2 Consider an automaton with the same β, α, and q as in Example 2.1. A stochastic state transition graph and stochastic output graph are shown in Figs. 2.5 and 2.6. The two mappings T and G are all stochastic. For the stochastic state transition function T, given the present state and input, various states, but not just one state, are likely to be reached. For the stochastic output function G, given the present state, various actions, but not just one action, are likely to be chosen. T can be defined in terms of conditional transition matrices

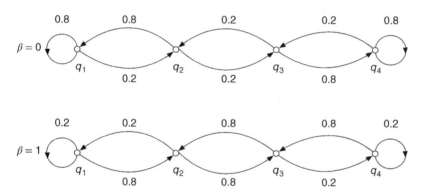

Figure 2.5 The stochastic state transition graph.

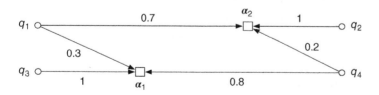

Figure 2.6 The stochastic output graph.

$T(\beta_0)$ and $T(\beta_1)$. Each of them is a $v \times v$ matrix following an input β. The elements in T have definition as follows:

$$\tau_{ij}^{\beta} = Pr\{q(t+1) = q_j | q(t) = q_i, \beta(t) = \beta\}. \quad i,j \in \mathbb{Z}_v, \beta \in B.$$

Similarly, G can be defined in terms of a matrix with dimension $v \times r$ whose elements are as follows:

$$g_{ij} = Pr\{\alpha(t) = \alpha_j | q(t) = q_i\}. \quad i \in \mathbb{Z}_v, j \in \mathbb{Z}_r.$$

According to Figs. 2.5 and 2.6, the corresponding stochastic state transition function T and the stochastic output function G can also be depicted with a matrix set and a matrix, respectively.

The stochastic state transition function T can be identified as

$$T(0) = \begin{array}{c} \\ q_1 \\ q_2 \\ q_3 \\ q_4 \end{array} \begin{array}{cccc} q_1 & q_2 & q_3 & q_4 \\ \left[\begin{array}{cccc} .8 & .2 & 0 & 0 \\ .8 & 0 & .2 & 0 \\ 0 & .2 & 0 & .8 \\ 0 & 0 & .2 & .8 \end{array}\right] \end{array} \quad T(1) = \begin{array}{c} \\ q_1 \\ q_2 \\ q_3 \\ q_4 \end{array} \begin{array}{cccc} q_1 & q_2 & q_3 & q_4 \\ \left[\begin{array}{cccc} .2 & .8 & 0 & 0 \\ .2 & 0 & .8 & 0 \\ 0 & .8 & 0 & .2 \\ 0 & 0 & .8 & .2 \end{array}\right] \end{array}.$$

The stochastic output function G can be identified as

$$G = \begin{array}{c} \\ q_1 \\ q_2 \\ q_3 \\ q_4 \end{array} \begin{array}{cc} \alpha_1 & \alpha_2 \\ \left[\begin{array}{cc} 0.3 & 0.7 \\ 0 & 1 \\ 1 & 0 \\ 0.8 & 0.2 \end{array}\right] \end{array}.$$

Assume that we have state q_1 and input $\beta = 0$ in the present instant. We have the transition probability to obtain the next state and may choose the next action as follows:

$$\tau_{11}^0 = Pr\{q(t+1) = q_1 | q(t) = q_1, \beta(t) = 0\} = 0.8$$
$$\tau_{12}^0 = Pr\{q(t+1) = q_2 | q(t) = q_1, \beta(t) = 0\} = 0.2$$
$$\tau_{13}^0 = Pr\{q(t+1) = q_3 | q(t) = q_1, \beta(t) = 0\} = 0$$
$$\tau_{14}^0 = Pr\{q(t+1) = q_4 | q(t) = q_1, \beta(t) = 0\} = 0$$

and

$$g_{11} = Pr\{\alpha(t) = \alpha_1 | q(t) = q_1\} = 0.3$$
$$g_{12} = Pr\{\alpha(t) = \alpha_2 | q(t) = q_1\} = 0.7$$

Apparently, in state q_1 and input $\beta = 0$, we have a chance to access different states and to choose different actions.

2.1.4 Measured Norms

The purpose of LAs is to interact with the environment to obtain as many rewards as possible from environmental feedbacks. For any two LAs, the one that receives more rewards is better than the one that receives fewer rewards. Therefore, LAs need to choose the "optimal behavior" to obtain the minimum penalty $\check{c}$. To measure LAs' performance, a series of mathematical terms are utilized to indicate the nature of LAs including *expedient, absolutely expedient, optimal,* and *ε-optimal.*

In the absence of prior information, each action follows the same distribution. LAs with this feature is called "pure chance automaton," which is usually used in various LAs. For the probability vector $P(t) = [p_1(t), p_2(t), ..., p_r(t)]$, the penalty mean is defined as follows:

$$
\begin{aligned}
\bar{\Phi} &= E[\beta(t)|P(t) = [p_1(t), p_2(t), ..., p_r(t)]] \\
&= \sum_{i=r} Pr[\beta(t) = 0|\alpha_i(t), P(t) = \{p_1(t), p_2(t), ..., p_r(t)\}] \cdot Pr[\alpha(t) = \alpha_i] \\
&= \sum_{i=r} Pr[\beta(t) = 0|\alpha_i(t)] \cdot Pr[\alpha(t) = \alpha_i] \\
&= \sum_{i=r} c_i \cdot p_i(t).
\end{aligned}
\tag{2.3}
$$

Thereby, for the pure chance automaton, the penalty mean is:

$$
\bar{\Phi}_0 = \sum_{i=r} c_i \cdot \frac{1}{r}.
\tag{2.4}
$$

An LA is *expedient* if it meets:

$$
\begin{aligned}
\lim_{t \to \infty} E\{\bar{\Phi}(t)\} &= \lim_{t \to \infty} E\{E\{\beta(t)|P(t)\}\} \\
&= \lim_{t \to \infty} E\{\beta(t)\} < \bar{\Phi}_0 \\
&= \sum_{i=r} c_i \cdot \frac{1}{r}.
\end{aligned}
\tag{2.5}
$$

It is *absolutely expedient* if it meets:

$$
E\{\bar{\Phi}(t+1)|P(t)\} < \bar{\Phi}(t).
\tag{2.6}
$$

If the expectations on both sides are separately computed, we can obtain:

$$
E\{E\{\bar{\Phi}(t+1)|P(t)\}\} < E\{\bar{\Phi}(t)\} \Leftrightarrow E\{\bar{\Phi}(t+1)\} < E\{\bar{\Phi}(t)\}.
\tag{2.7}
$$

This means that $E\{\bar{\Phi}(t)\}$ is strictly monotonously decreasing over time.

An LA is *optimal* if the probability of an action with the least penalty satisfies:

$$
\lim_{t \to \infty} P_{min}(t) \to 1, w.p.1,
\tag{2.8}
$$

where "w.p.1" means with probability 1.

In fact, the optimality condition mentioned above is so strict that many LAs cannot satisfy it. Therefore, a slightly weaker definition called ε-optimality is proposed. An LA is ε-optimal, and if for each $\varepsilon > 0$ and $\delta > 0$, there is an instant $t_0 < \infty$ and a learning parameter $\lambda_0 > 0$, such that for all $t \geq t_0$ and $\lambda < \lambda_0$:

$$Pr\{|p_{min}(t) - 1| < \varepsilon\} > 1 - \delta. \tag{2.9}$$

In the sense of penalty mean, this definition with any $\varepsilon > 0$ can also be written as:

$$\lim_{t \to \infty} E\{\bar{\Phi}(t)\} < \check{c} + \varepsilon. \tag{2.10}$$

Intuitively, ε-optimality means that if given enough time and appropriate parameters, the probability of an LA to choose the optimal action is infinitely close to 1.

2.2 Fixed Structure Learning Automata

In the last section, we have introduced a deterministic and stochastic LA. For different automata, we can make the following summary based on their different characteristics: In a deterministic automaton, the elements in the state transition matrices are composed of 1 and 0 only. Therefore, given an input, the state transition process is deterministic, and it is thus called a deterministic automaton. In a fixed-structure stochastic automaton, the elements in the state transition matrices are composed of values in the interval [0, 1]. Therefore, given an input, the state transition process is indeterminate and therefore belongs to a stochastic automaton. The elements in the state transition matrices are fixed, and such stochastic automaton is thus called a fixed structure stochastic automaton. In a variable-structure stochastic automaton, the elements in the state transition matrices are composed of values in the interval [0, 1]. Therefore, given an input, the state transition process is indeterminate, therefore leading to a stochastic automaton. The elements in the state transition matrices are variable, and such stochastic automaton is thus called a variable-structure stochastic automaton.

There are several common fixed-structure automata, i.e., Tsetlin [1], Krylov [2], Krinsky [2], IJA (Iraji–Jamalian Automaton) [3], and TFSLA (Tunable Fixed Structure Learning Automaton) [4]. We introduce them next.

2.2.1 Tsetlin Learning Automaton

A Tsetlin automaton is the earliest proposed LA. Its state transition graph is shown in Fig. 2.7.

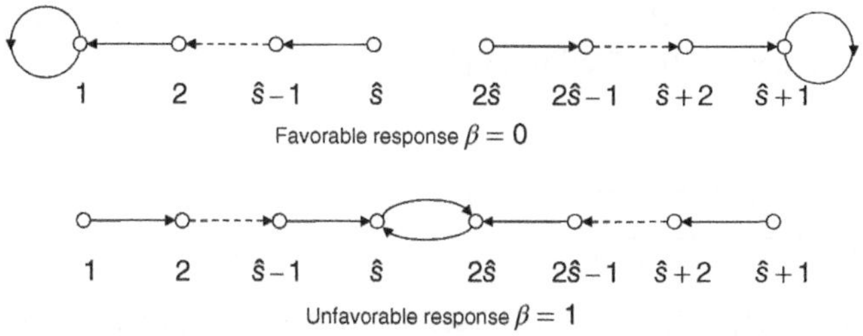

Favorable response $\beta = 0$

Unfavorable response $\beta = 1$

Figure 2.7 The state transition graph of Tsetlin LA.

In Tsetlin LA, there are two actions, $2\hat{s}$ states and each action can lead a state to one of $\hat{s}$ states. This LA is denoted as $\bar{L}_{2\hat{s},2}$. It is easy to expand to r actions with $r \cdot \hat{s}$ states.

The output function of a Tsetlin LA is relatively simple. When it is in state $q(t) = q_i$, $i \in Z_{\hat{s}}$, the output is α_1. If it is in state $q(t) = q_i$, $i \in Z_{2\hat{s}} - Z_{\hat{s}}$, the output is α_2. The output function can be defined as:

$$G(q_i) = \begin{cases} \alpha_1, i \in Z_{\hat{s}} \\ \alpha_2, i \in Z_{2\hat{s}} - Z_{\hat{s}} \end{cases}.$$

The state transition following an action can be expressed as:

$$\left. \begin{array}{l} q_i \rightarrow q_{i+1} \quad (i \in Z_{2\hat{s}-1} - Z_{\hat{s}}) \\ q_{2\hat{s}} \rightarrow q_{\hat{s}} \end{array} \right\} \alpha_2 \text{ results in an unfavorable response,}$$

$$\left. \begin{array}{l} q_i \rightarrow q_{i-1} \quad (i \in Z_{2\hat{s}-1} - Z_{\hat{s}}) \\ q_{\hat{s}+1} \rightarrow q_{\hat{s}+1} \end{array} \right\} \alpha_2 \text{ results in a favorable response.}$$

$$(2.11)$$

$$\left. \begin{array}{l} q_i \rightarrow q_{i+1} \quad (i \in Z_{\hat{s}-1}) \\ q_{\hat{s}} \rightarrow q_{2\hat{s}} \end{array} \right\} \alpha_1 \text{ results in an unfavorable response,}$$

$$\left. \begin{array}{l} q_i \rightarrow q_{i-1} \quad (i \in Z_{\hat{s}} - 1) \\ q_1 \rightarrow q_1 \end{array} \right\} \alpha_1 \text{ results in a favorable response.}$$

$$(2.12)$$

At the initial moment, a Tsetlin LA is randomly in state $q_{2\hat{s}}$ or $q_{\hat{s}}$ without any priori knowledge. After obtaining a feedback from the environment, the state transition is performed according to the state transition graph mentioned in Fig. 2.7. For example, when $q(t) = q_{\hat{s}}$, its output action is α_1. Then if the Tsetlin LA receives the reward feedback from the environment, the state changes to $q(t) = q_{\hat{s}-1}$; on the contrary, if the feedback obtained from the environment is penalty, then the state is transferred to $q(t) = q_{2\hat{s}}$.

The finite state irreducible Markov chain is ergodic. So is the LA. It has been proved that the penalty mean of the Tsetlin LA [1] is:

$$\bar{\Phi}(\bar{L}_{2\hat{s},2}) = \frac{\frac{1}{\check{\beta}_1^{\hat{s}-1}} \cdot \frac{\hat{\beta}_1^{\hat{s}} - \check{\beta}_1^{\hat{s}}}{\hat{\beta}_1 - \check{\beta}_1} + \frac{1}{\check{\beta}_2^{\hat{s}-1}} \cdot \frac{\hat{\beta}_2^{\hat{s}} - \check{\beta}_2^{\hat{s}}}{\hat{\beta}_2 - \check{\beta}_2}}{\frac{1}{\check{\beta}_1^{\hat{s}}} \cdot \frac{\hat{\beta}_1^{\hat{s}} - \check{\beta}_1^{\hat{s}}}{\hat{\beta}_1 - \check{\beta}_1} + \frac{1}{\check{\beta}_2^{\hat{s}}} \cdot \frac{\hat{\beta}_2^{\hat{s}} - \check{\beta}_2^{\hat{s}}}{\hat{\beta}_2 - \check{\beta}_2}}, \check{\beta}_1 + \hat{\beta}_1 = 1, \check{\beta}_2 + \hat{\beta}_2 = 1. \tag{2.13}$$

In the environment $min\{\check{\beta}_1, \check{\beta}_2\} < 0.5$, the Tsetlin LA has been proven to be ε-optimal [5].

2.2.2 Krinsky Learning Automaton

Similar to Tsetlin LA, in Krinsky LA, there are two actions, $2\hat{s}$ states and each action can lead a state to one of $\hat{s}$ states. Krinsky LA is denoted as $\bar{K}_{2\hat{s},2}$. Its output function is the same as Tsetlin LAs. When it is in state $q(t) = q_i$, $i \in \mathbb{Z}_{\hat{s}}$, its output is α_1. If it is in state $q(t) = q_i$, $i \in \mathbb{Z}_{2\hat{s}} - \mathbb{Z}_{\hat{s}}$, its output is α_2. Figure 2.8 shows the state transition graph of the Krinsky LA. The output function can be defined as:

$$\left.\begin{array}{l} q_i \to q_{i+1} \quad (i \in \mathbb{Z}_{2\hat{s}-1} - \mathbb{Z}_{\hat{s}}) \\ q_{2\hat{s}} \to q_{\hat{s}} \end{array}\right\}, \alpha_2 \text{ results in an unfavorable response,}$$
$$q_i \to q_{\hat{s}+1}, \quad \alpha_2 \text{ results in a favorable response.}$$

$$\tag{2.14}$$

$$\left.\begin{array}{l} q_i \to q_{i+1} \quad (i \in \mathbb{Z}_{\hat{s}-1}) \\ q_{\hat{s}} \to q_{2\hat{s}} \end{array}\right\}, \alpha_1 \text{ results in an unfavorable response,} \tag{2.15}$$
$$q_i \to q_1, \quad \alpha_1 \text{ results in a favorable response.}$$

At the initial moment, the Krinsky LA is randomly in state $q_{2\hat{s}}$ or $q_{\hat{s}}$ without any priori knowledge. When getting an environmental penalty, its state transition process is the same as Tsetlin LAs. In Krinsky LA, each state jumps to the deepest state when the environment rewards the current action as shown in Fig. 2.8, which differs from Tsetlin LA's behavior. For example, when Krinsky LA is in state $q_{\hat{s}}$, the output action is α_1. Then if the feedback from the environment is a penalty, the state is transferred to $q_{2\hat{s}}$, which is the same as the Tsetlin LA. On the contrary, if the feedback obtained from the environment is a reward, the state is transferred to q_1, while Tsetlin LA transitions its state to $q_{\hat{s}-1}$. The penalty mean of the Krinsky LA is:

$$\bar{\Phi}(\bar{K}_{2\hat{s},2}) = \frac{\frac{1}{\check{\beta}_1^{\hat{s}-1}} + \frac{1}{\check{\beta}_2^{\hat{s}-1}}}{\frac{1}{\check{\beta}_1^{\hat{s}}} + \frac{1}{\check{\beta}_2^{\hat{s}}}}, \check{\beta}_1 + \hat{\beta}_1 = 1, \check{\beta}_2 + \hat{\beta}_2 = 1 \tag{2.16}$$

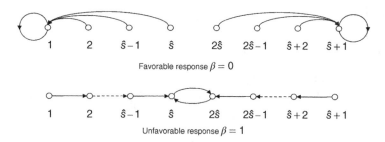

Figure 2.8 The state transition graph of Krinsky LA.

The Krinsky LA has been proven to be ε-optimal [5] in any deterministic environment.

2.2.3 Krylov Learning Automaton

Similar to Tsetlin LA, in Krylov LA, there are two actions, $2\hat{s}$ states and each action can lead a state to one of $\hat{s}$ states. Krylov LA is denoted as $\bar{J}_{2\hat{s},2}$. Its output function is the same as Tsetlin LA's. When it is in state $q(t) = q_i$, $i \in \mathbb{Z}_{\hat{s}}$, its output is α_1. If it is in state $q(t) = q_i$, $i \in \mathbb{Z}_{2\hat{s}} - \mathbb{Z}_{\hat{s}}$, its output is α_2. Figure 2.9 shows the state transition graph of the Krylov LA.

At the initial moment, the Krylov LA is randomly in state $q_{2\hat{s}}$ or $q_{\hat{s}}$ without any priori knowledge. When getting an environmental reward, the state transition process in Krylov is the same as in Tsetlin LA. Differently from the Tsetlin LA, in Krylov LA, each state transfers to the two adjacent states with the probability of 0.5 when the environment rewards the current action. For example, when Krylov LA is in state $q_{\hat{s}}$, the output action is α_1. Then if the feedback from the environment is a reward, the state is transferred to $q_{\hat{s}-1}$, which is the same as in the Tsetlin LA. On the contrary, if the feedback obtained from the environment

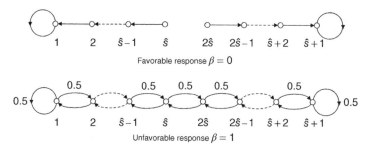

Figure 2.9 The state transition graph of krylov LA.

is a penalty, the state is transferred to $q_{\hat{s}-1}$ or $q_{2\hat{s}}$ with a same probability of 0.5.

The penalty mean of the Krylov LA is:

$$\bar{\Phi}(\bar{J}_{2\hat{s},2}) = \frac{\frac{1}{\check{\beta}_1^{\hat{s}-1}} \cdot \frac{\check{\beta}_1^{\hat{s}}-1}{\check{\beta}_1-1} + \frac{1}{\check{\beta}_2^{\hat{s}-1}} \cdot \frac{\check{\beta}_2^{\hat{s}}-1}{\check{\beta}_2-1}}{\frac{1}{\tilde{\beta}_1} \cdot \frac{1}{\check{\beta}_1^{\hat{s}-1}} \cdot \frac{\check{\beta}_1^{\hat{s}}-1}{\check{\beta}_1-1} + \frac{1}{\tilde{\beta}_2} \cdot \frac{1}{\check{\beta}_2^{\hat{s}-1}} \cdot \frac{\check{\beta}_2^{\hat{s}}-1}{\lambda_2-1}}, \tilde{\beta}_1 = \frac{1}{1-\check{\beta}_1}, \tilde{\beta}_2 = \frac{1}{1-\check{\beta}_2} \quad (2.17)$$

The limit of the penalty mean is: $\lim_{\hat{s}\to\infty}\bar{\Phi}(\bar{J}_{2\hat{s},2}) = min\{\check{\beta}_1, \check{\beta}_2\}$, implying that the Krylov LA is ε-optimal [5] in any deterministic environment.

2.2.4 IJA Learning Automaton

In IJA LA (Iraji–Jamalian Automaton), there are two actions and $2\hat{s}$ states. Each action can lead a state to one of $\hat{s}$ states. IJA LA is denoted as $\bar{I}_{2\hat{s},2}$. Figure 2.10 shows the state transition graph of an IJA LA. We use symbols $\hat{\beta}_1$ to represent favorable responses and $\check{\beta}_2$ to represent unfavorable ones. Its output function is the same as Tsetlin LAs.

At the initial moment, IJA LA is randomly in state $q_{2\hat{s}}$ or $q_{\hat{s}}$ without any priori knowledge. When receiving an environmental penalty, its state transition process in IJA is the same as Tsetlin LAs. Differently from Tsetlin LA, in IJA LA, when receiving an environmental reward, its state is transferred to $q_{\left[\frac{\hat{s}}{2}\right]}$ with the probability of 0.5.

The penalty mean of IJA LA is:

$$\bar{\Phi}(\bar{I}_{2\hat{s},2}) = \frac{\check{\beta}_1\tilde{I}_1\tilde{I}_2 + \check{\beta}_2\tilde{I}_3\tilde{I}_4}{\tilde{I}_1\tilde{I}_2 + \tilde{I}_3\tilde{I}_4}. \quad (2.18)$$

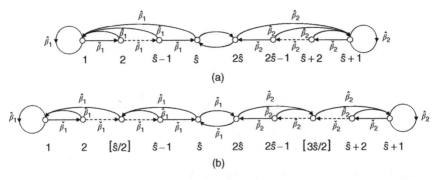

Figure 2.10 The state transition graph of IJA LA. (a) $\hat{\beta}_1$ represent favorable response and (b) $\check{\beta}_2$ represent unfavorable one.

The relevant parameters are as follows:

$$\tilde{I}_4 = \frac{\check{\beta}_1^{\lfloor \frac{3}{2} \rfloor}}{\frac{3}{2}(1-\check{\beta}_1)-1}, \tilde{I}_3 = \frac{1-\check{\beta}_2^3}{1-\check{\beta}_2} + \frac{\hat{s} \cdot \check{\beta}_2^{\lfloor \frac{3}{2} \rfloor}}{\frac{3}{2}(1-\check{\beta}_2)-1},$$

$$\tilde{I}_2 = \frac{\check{\beta}_2^3}{1-\check{\beta}_1} + \frac{\check{\beta}_2^{\lfloor \frac{3}{2} \rfloor+1}}{(1-\check{\beta}_1)\left[\frac{3}{2}(1-\check{\beta}_2)-1\right]}, \tilde{I}_1 = \frac{1-\check{\beta}_1^3}{1-\check{\beta}_1} + \frac{\hat{s} \cdot c_1^{\lfloor \frac{3}{2} \rfloor}}{\frac{3}{2}(1-\check{\beta}_1)-1}, \quad (2.19)$$

The limit of the penalty mean is: $\lim_{\hat{s} \to \infty} \bar{\Phi}(\tilde{I}_{2\hat{s},2}) = min\{\check{\beta}_1, \check{\beta}_2\}$, which means IJA LA is ε-optimal [3] in any deterministic environment.

2.3 Variable Structure Learning Automata

In Section 2.2, we have introduced the fixed structure stochastic automata in stationary random environments. Their state transition probabilities and action probabilities are fixed.

In a pioneering paper, Varshavskii and Vorontsova first present automata that update transition probabilities [6]. Fu and his associates give an extension updating action probabilities [7–9]. In this book, the emphasis is on schemes for updating action probabilities.

Compared with the fixed structure stochastic automata, the variable structure learning automata could change its state with iteration count according to the environment, and has the advantages of faster convergence and adaptive learning ability. Its quintuple $\langle A, B, Q, T, G \rangle$ model can be simplified into a quadruple $\langle A, B, Q, T \rangle$. The concepts of A, B, and T have the same definitions as introduced before. In variable structure LAs, $Q = \langle P, \mathbb{E} \rangle$. Here, P represents the action probability matrix, and $\mathbb{E}$ is the estimator. Because the state is composed of two parts, the update of the state is also divided into two parts: updates P and update $\mathbb{E}$. The updated formulas for the estimator and probability matrix are expressed as

$$\mathbb{E}(t+1) = T_{\mathbb{E}}(\mathbb{E}(t), \alpha(t), \beta(t)) \quad (2.20)$$

$$P(t+1) = T_P(P(t), \alpha(t), \beta(t), \mathbb{E}(t+1)). \quad (2.21)$$

Therefore, the state transition function should also consist of two parts: $T = \langle T_{\mathbb{E}}, T_P \rangle$.

From the updated formula of the probability vector, we know that $\{P(t)\}_{t \geq 0}$ is a discrete time homogeneous Markov process. According to whether the Markov process is ergodic or absorbing, the update algorithm can be referred to as an ergodic algorithm and an absorbing one [10–12]. If the updated formula of $P(t+1) = T_P(P(t), \alpha(t), \beta(t), \mathbb{E}(t+1))$ is linear, such LAs are called linear variable structure LA; otherwise, nonlinear variable structure LA.

In the update strategies, there are two basic principles for LA algorithms: when the output behavior is punished by the environment, the probability of this behavior should be reduced; otherwise, if the output behavior is rewarded by the environment, the probability of this behavior should be increased. Different specific algorithms may use different principles. Examples of combinations of these principles are as follows:

1) RP (Reward Penalty): The probability vector is updated when it is rewarded and punished by the environment;
2) RI (Reward Inaction): The probability vector is only updated when it is rewarded by the environment but not when receiving an environmental penalty; and
3) IP (Inaction Penalty): The probability vector is updated only when it is punished by the environment. The probability vector does not change when receiving environmental reward.

Variable structure learning automata can be further divided into three categories.

1) The first type is an estimator-free LA, which is specifically represented by a quadruple $\langle A, B, Q, T \rangle$, where $Q = \langle P, \mathbb{E} \rangle$ and $\mathbb{E} = \emptyset$;
2) The second type is LA with a deterministic estimator. This type of algorithms is represented by a Pursuit estimator framework, which is specifically represented by the a quadruple $\langle A, B, Q, T \rangle$, where $Q = \langle P, \mathbb{E} \rangle$, $\mathbb{E} = \tilde{D}$ and $\tilde{D}$ is a deterministic estimator; and
3) The third type is an automatic learning machine with a stochastic estimator represented by the SE_{RI} algorithm, which is specifically represented by a quadruple $\langle A, B, Q, T \rangle$, where $Q = \langle P, \mathbb{E} \rangle$, $\mathbb{E} = \langle \tilde{D}, \tilde{S} \rangle$, and $\tilde{D}$ and $\tilde{S}$ are denoted as deterministic and stochastic estimators, respectively.

2.3.1 Estimator-Free Learning Automaton

The work [6] is the first to propose an estimator-free LA. The LA update algorithms proposed earlier are all calculated on a probability space that is continuous. To be precise, the probability of each behavior can be any value in the interval $[0, 1]$. Oommen et al. discretize the continuous probability space and introduce discrete space LA. Applying the three strategies RP, RI, and IP mentioned above to continuous and discrete linear formulas, six different linear strategies can be obtained: L_{RP}, L_{RI}, L_{IP}, DL_{RP}, DL_{RI}, and DL_{IP} [10, 12, 13]. In this section, we introduce the typical LAs with these linear strategies.

The algorithms introduced below are based on the assumption that there are only two behaviors denoted as a set $A = \{\alpha_1, \alpha_2\}$, and these algorithms can also be extended to the case of multiple behaviors $A = \{\alpha_1, \alpha_2, ..., \alpha_r\}$. Then the probability space P is two-dimensional, i.e., $P(t) = [p_1(t), p_2(t)]$.

1) Continuous Linear Schemes: L_{RP}, L_{RI}, and L_{IP}.

L_{RP} means that an LA directly increases the action probability linearly when it receives a reward feedback from the environment, and decreases the action probability when it receives a penalty feedback from the environment. L_{RI} means that LA increases the action probability linearly when it receives a reward feedback from the environment, but does not change the probability of the action when it receives a penalty feedback from the environment. L_{IP} means that LA decreases the action probability linearly when it receives a penalty feedback from the environment, but does not change the probability of the action when it receives a reward feedback from the environment.

The probability vector update formula of L_{RP}, L_{RI}, and L_{IP} is described as follows where parameters $\lambda_1 \in (0, 1)$ and $\lambda_2 \in (0, 1)$ are the reward and penalty parameters, respectively; and at any instant t, $p_1(t) + p_2(t) = 1$.

$$
\begin{aligned}
&p_1(t+1) = p_1(t) + \lambda_1 \cdot (1 - p_1(t)) &&\text{if } \alpha(t) = \alpha_1 \text{ and } \beta_t = 1\\
&p_1(t+1) = (1 - \lambda_2) \cdot p_1(t) &&\text{if } \alpha(t) = \alpha_1 \text{ and } \beta_t = 0\\
&p_1(t+1) = (1 - \lambda_1) \cdot p_1(t) &&\text{if } \alpha(t) = \alpha_2 \text{ and } \beta_t = 1\\
&p_1(t+1) = p_1(t) + \lambda_2 \cdot (1 - p_1(t)) &&\text{if } \alpha(t) = \alpha_2 \text{ and } \beta_t = 0
\end{aligned}
\tag{2.22}
$$

$$
\begin{aligned}
&p_1(t+1) = p_1(t) + \lambda_1 \cdot (1 - p_1(t)) &&\text{if } \alpha(t) = \alpha_1 \text{ and } \beta_t = 1\\
&p_1(t+1) = p_1(t) &&\text{if } \alpha(t) = \alpha_1 \text{ and } \beta_t = 0\\
&p_1(t+1) = (1 - \lambda_1) \cdot p_1(t) &&\text{if } \alpha(t) = \alpha_2 \text{ and } \beta_t = 1\\
&p_1(t+1) = p_1(t) &&\text{if } \alpha(t) = \alpha_2 \text{ and } \beta_t = 0
\end{aligned}
\tag{2.23}
$$

$$
\begin{aligned}
&p_1(t+1) = p_1(t) &&\text{if } \alpha(t) = \alpha_1 \text{ and } \beta_t = 1\\
&p_1(t+1) = (1 - \lambda_2) \cdot p_1(t) &&\text{if } \alpha(t) = \alpha_1 \text{ and } \beta_t = 0\\
&p_1(t+1) = p_1(t) &&\text{if } \alpha(t) = \alpha_2 \text{ and } \beta_t = 1\\
&p_1(t+1) = p_1(t) + \lambda_2 \cdot (1 - p_1(t)) &&\text{if } \alpha(t) = \alpha_2 \text{ and } \beta_t = 0.
\end{aligned}
\tag{2.24}
$$

It is worth noting that the parameters satisfy $\lambda_1 \in (0, 1)$ and $\lambda_2 \in (0, 1)$. When $\lambda_1 \in (0, 1)$ and $\lambda_2 = 0$, L_{RP} becomes L_{RI}. When $\lambda_2 \in (0, 1)$ and $\lambda_1 = 0$, L_{RP} becomes L_{IP}.

The above three types of LAs L_{RP}, L_{RI}, and L_{IP} work in continuous probability space. In order to accelerate their convergence speed, researchers have proposed discrete space learning automata algorithms.

2) Discrete Linear Schemes: DL_{RP}, DL_{RI}, and DL_{IP}.

The basic idea of discretization is to divide the area [0, 1] into Δ parts, and each part is $1/\Delta$ in length, so the probability value that can be taken becomes a finite set $\{0, 1/\Delta, 2/\Delta, ..., (\Delta - 1)/\Delta, 1\}$. Its essence is to transform a homogeneous Markov process into a homogeneous Markov chain.

The probability vector update formula of DL_{RP}, DL_{RI}, and DL_{IP} described as follows where at any instant t, $p_1(t) + p_2(t) = 1$.

if $0 < p_1(t) < 1$,

$$
\begin{aligned}
p_1(t+1) &= p_1(t) + \tfrac{1}{\Delta} && \text{if } \alpha(t) = \alpha_1 \text{ and } \beta_t = 1 \\
p_1(t+1) &= \left(1 - \tfrac{1}{\Delta}\right) && \text{if } \alpha(t) = \alpha_1 \text{ and } \beta_t = 0 \\
p_1(t+1) &= p_1(t) + \tfrac{1}{\Delta} && \text{if } \alpha(t) = \alpha_2 \text{ and } \beta_t = 0 \\
p_1(t+1) &= \left(1 - \tfrac{1}{\Delta}\right) && \text{if } \alpha(t) = \alpha_2 \text{ and } \beta_t = 1
\end{aligned}
\tag{2.25}
$$

if $p_1(t) \in \{0,1\}$,

$$
\begin{aligned}
p_1(t+1) &= p_1(t) && \text{if } p_1(t) \in \{0,1\} \text{ and } \beta_t = 1 \\
p_1(t+1) &= \tfrac{1}{\Delta} && \text{if } p_1(t) = 0 \text{ and } \beta_t = 0 \\
p_1(t+1) &= \left(1 - \tfrac{1}{\Delta}\right) && \text{if } p_1(t) = 1 \text{ and } \beta_t = 0
\end{aligned}
$$

$$
\begin{aligned}
p_1(t+1) &= \min\left\{p_1(t) + \tfrac{1}{\Delta}, 1\right\} && \text{if } \alpha(t) = \alpha_1 \text{ and } \beta_t = 1 \\
p_1(t+1) &= p_1(t) && \text{if } \alpha(t) = \alpha_1 \text{ and } \beta_t = 0 \\
p_1(t+1) &= \max\left\{p_1(t) - \tfrac{1}{\Delta}, 0\right\} && \text{if } \alpha(t) = \alpha_2 \text{ and } \beta_t = 1 \\
p_1(t+1) &= p_1(t) && \text{if } \alpha(t) = \alpha_2 \text{ and } \beta_t = 0
\end{aligned}
\tag{2.26}
$$

$$
\begin{aligned}
p_1(t+1) &= \max\left\{p_1(t) - \tfrac{1}{\Delta}, 0\right\} && \text{if } \alpha(t) = \alpha_1 \text{ and } \beta_t = 0 \\
p_1(t+1) &= p_1(t) && \text{if } \alpha(t) = \alpha_1 \text{ and } \beta_t = 1 \\
p_1(t+1) &= \min\left\{p_1(t) + \tfrac{1}{\Delta}, 0\right\} && \text{if } \alpha(t) = \alpha_2 \text{ and } \beta_t = 0 \\
p_1(t+1) &= p_1(t) && \text{if } \alpha(t) = \alpha_2 \text{ and } \beta_t = 1
\end{aligned}
\tag{2.27}
$$

2.3.2 Deterministic Estimator Learning Automaton

To improve the convergence speed of LA algorithms, Thathachar and Sastry [14–16] add an estimator to the original algorithm. The estimator-based algorithm introduces an estimate of the environmental reward probability, and uses this estimate to update the probability vector. Depending on whether the estimator is deterministic or random, it can be divided into deterministic and stochastic estimators. In this subsection we introduce the deterministic estimator algorithm. Moreover, according to different update methods of the probability matrix, the deterministic estimator algorithm can be divided into a continuous deterministic estimator algorithm and a discrete one [17, 18]. The deterministic estimator algorithm is defined as a quaternion $\langle A, B, Q, T \rangle$, where

1) $A = \{\alpha_1, \alpha_2, ..., \alpha_r\}$ is a set of actions ($2 \le r < \infty$).
2) $B = \{0, 1\}$. The environmental response at instant t is denoted by $\beta(t)$. "1" and "0" denote the reward and penalty responses, respectively.

3) $Q = \langle P, \mathbb{E} \rangle$ is the state set. $P = \{p_1(t), p_2(t), ..., p_r(t), \}$ is the state of the automaton at instant t, where $p_i(t) = Pr\{\alpha(t) = \alpha_i\}$. $\mathbb{E}$ is the estimator, and the deterministic estimator is defined as $\mathbb{E} = \tilde{D}(t)$. $\tilde{D}(t) = \{\tilde{d}_1(t), \tilde{d}_2(t), ... \tilde{d}_r(t)\}$ is the estimator vector at instant t. The estimated reward for each behavior is:

$$\tilde{d}_i(t) = \frac{\mathbb{H}_i(t)}{\mathbb{G}_i(t)},$$

where $\mathbb{H}_i(t)$ represents the number of times that the ith action has been rewarded up to instant t, $i \in \mathbb{Z}_r$. $\mathbb{G}_i(t)$ is the number of times that the ith action has been selected up to instant t, $i \in \mathbb{Z}_r$.

4) $T : Q \times B \rightarrow Q$ is the state transfer function of the automaton. T determines how the automaton migrates to the state at $t + 1$ according to the output, input and the state at instant t.

In the estimator-free algorithm, the probability vector is updated based on the selected behavior and feedback from the environment. In the estimator algorithm, the estimator is updated according to the selected behavior and environment feedback information, and finally the information of the estimator is used to update the probability vector. In order to more intuitively reflect the steps of the estimator algorithm, the continuous pursuit reward-penalty algorithm is given next.

CP_{RP}: **Continuous Pursuit Reward–Penalty Algorithm**

Parameters and Notation

λ: learning parameter where $\lambda \in (0, 1)$.

m: index of the maximal component of $\tilde{D}(t)$, $\tilde{d}_m(t) = max_i\{\tilde{d}_i(t)\}$.

$\hat{T}$: threshold value.

$\mathbb{H}_i(t)$: number of times the ith action has been rewarded up to instant t, $i \in \mathbb{Z}_r$.

$\mathbb{G}_i(t)$: number of times the ith action has been selected up to instant t, $i \in \mathbb{Z}_r$.

Input: Number of allowable actions r, learning parameter λ, action set A, environmental response set B, convergence threshold $\hat{T}$.

Output: Estimated optimal action α_m.

Initialize $p_i(0) = \frac{1}{r}, i \in \mathbb{Z}_r$

Initialize $\tilde{D}(0)$ by picking each action a small number of times.

do

1. At instant t, pick an action $\alpha(t)$ according to the probability distribution $P(t)$.

2. Obtain feedback $\beta(t)$ from the environment according to behavior $\alpha(t)$.

3. Update $\tilde{D}(t)$ by following:

$\mathbb{G}_i(t) = \mathbb{G}_i(t) + 1$;

$\mathbb{H}_i(t) = \mathbb{H}_i(t) + \beta(t)$;

$\tilde{d}_i(t) = \frac{\mathbb{H}_i(t)}{\mathbb{G}_i(t)}.$

4. Update $P(t)$ by following:

$$p_m(t+1) = p_m(t) + \lambda \cdot (1 - p_m(t));$$
$$p_j(t+1) = p_j(t) + \lambda \cdot (1 - p_j(t)), j \neq m.$$

while$(max\{p_j(t+1)\} \leq \hat{T})$

END

2.3.3 Stochastic Estimator Learning Automaton

Stochastic estimator LAs, as the name implies, adopt stochastic estimators. SE_{RI} [19] is a classic stochastic estimator algorithm that has been proven to be ε-optimal. Therefore, in this section, we utilize the SE_{RI} algorithm to introduce stochastic estimator LAs. SE_{RI} is defined as a quaternion $\langle A, B, Q, T \rangle$.

1) $A = \{\alpha_1, \alpha_2, ..., \alpha_r\}$ is a set of actions ($2 \leq r < \infty$).
2) $B = \{0, 1\}$. The environmental response at instant t is denoted by $\beta(t)$.
3) $Q = \langle P, \mathbb{E} \rangle$ is the state set. $P = \{p_1(t), p_2(t), ..., p_r(t)\}$ is the state of the automaton at instant t, where $p_i(t) = Pr\{\alpha(t) = \alpha_i\}$. $\mathbb{E}$ is the estimator, and it is defined as $\mathbb{E} = (\tilde{D}(t), \tilde{S}(t))$. $\tilde{D}(t) = \{\tilde{d}_1(t), \tilde{d}_2(t), ... \tilde{d}_r(t)\}$ is the same as a deterministic estimator's. $\tilde{S}(t) = \{\tilde{s}_1(t), \tilde{s}_2(t), ... \tilde{s}_r(t)\}$ is a stochastic estimator vector at instant t, and it is defined as:

$$\tilde{s}_i(t) = \tilde{d}_i(t) + Z_i(t),$$

where $Z_i(t)$ is a random variable that is uniformly distributed within $(\frac{-\gamma}{G_i(t)}, \frac{\gamma}{G_i(t)})$.

The SE_{RI} algorithm is given as follows.

SE_{RI}: **Stochastic Estimator Reward–Inaction Algorithm**

Parameters and Notation

m: index of the maximal component of $\tilde{S}(t)$, $\tilde{s}_m(t) = max_i\{\tilde{s}_i(t)\}$.

γ: perturbation parameter.

$\hat{T}$: threshold value.

Δ: step size.

$\mathbb{H}_i(t)$: number of times the ith action has been rewarded up to instant t, $i \in \mathbb{Z}_r$.

$\mathbb{G}_i(t)$: number of times the ith action has been selected up to instant t, $i \in \mathbb{Z}_r$.

Input: Number of allowable actions r, learning parameter λ, action set A, environmental response set B, perturbation parameter γ, convergence threshold $\hat{T}$.

Output: Estimated optimal action α_m.

Initialize $p_i(0) = \frac{1}{r}, i \in \mathbb{Z}_r$

Initialize $\tilde{D}(0)$ by picking each action a small number of times.

do

    1: At instant t pick an action $\alpha(t)$ according to the probability distribution $P(t)$.

    2: Get feedback $\beta(t)$ from the environment according to behavior $\alpha(t)$.

3: Update $\widetilde{D}(t)$ by following:

$\mathbb{G}_i(t) = \mathbb{G}_i(t) + 1;$

$\mathbb{H}_i(t) = \mathbb{H}_i(t) + \beta(t);$

$\widetilde{d}_i(t) = \frac{\mathbb{H}_i(t)}{\mathbb{G}_i(t)};$

$\widetilde{s}_i(t) = \widetilde{d}_i(t) + Z_i(t);$

$\widetilde{s}_m(t) = max_i\{\widetilde{s}_i(t)\}.$

4: Update $P(t)$ by following:

if $\beta(t) = 1, \begin{cases} p_j(t+1) = max\{p_j(t) - \Delta, 0\} & j \neq m \\ p_m(t+1) = 1 - \sum_{j \neq m} p_j(t+1) \end{cases}$.

else $p_j(t+1) = p_j(t), j \in \mathbb{Z}_r$

while$(max\{p_j(t+1)\} \leq \hat{T})$

END

2.4 Summary

In this chapter, we first introduce an LA and its components, namely automata and environment, and how it works. Then, mathematical terms for describing its nature are introduced. According to its development, LAs can be classified. In terms of whether the state transition and output functions of an automaton are deterministic or stochastic, LAs can be divided into deterministic and stochastic automata. Compared with the former, the latter can adapt to environmental conditions to change its states, converge faster, and achieve the purpose of adaptive learning. These advantages make stochastic automata more widely studied and used [20–24]. According to whether the state transition and output functions of an automaton change with time, the stochastic automaton can be divided into fixed and variable structure LAs. In terms of whether there is an estimator and whether the estimator is stochastic, LAs with a variable structure can be divided into estimator-free, deterministic estimator, and stochastic estimator LAs. This chapter depicts the learning automata with these different characteristics in detail. Figure 2.11 illustrates the LA classification.

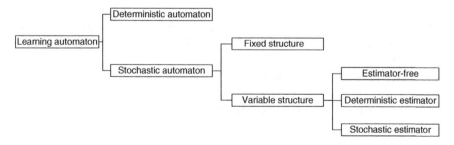

Figure 2.11 The classification of LA.

2.5 Exercises

1 What is the mathematical model of learning automata?

2 Please list several examples where learning automata have been applied.

3 What are the components of an automaton?

4 What is the difference among P-model, Q-model, and S-model random environments.

5 Please express how an automaton interacts with the environment and draw an interaction diagram.

6 What is the difference between deterministic and stochastic automata? Make an example for each of the two automata and draw their corresponding Markov state transition processes.

7 How to measure the performance of an LA?

8 The probability update formula listed in (2.22) ~ (2.24) are formulas for two actions. Please extend them to the case of r actions.

9 What is the difference between the fixed structure stochastic LA and the variable structure stochastic LA?

10 Please draw the state transition graphs of Tsetlin LA, Krinsky LA, Krylov LA, and IJA LA.

11 According to the automaton performance listed in (2.5) ~ (2.10), please prove that Krinsky LA is ε-optimal in any deterministic environment.

12 Please list the common combinations of the action probabilities update principles. What is the difference among them?

13 Please classify variable structure LAs. What are their estimators?

14 What is the main difference between estimator-free and deterministic estimator LAs?

15 In Section 2.3.2, the basic continuous pursuit rewardâ–penalty algorithm (CP_{RP}) is presented. Based on the definitions of RP, RI, and IP mechanisms and CP_{RP}, please write the pseudo code of discrete reward inaction pursuit algorithm: DP_{RI}.

References

1 M. L. Tsetlin, "On the behavior of finite automata in random media," *Automatika i Telemekhanika*, vol. 22, pp. 1345–1354, Oct 1961.

2 M. L. Tsetlin, "Automaton theory and the modeling of biological systems," New York: Academic, 1973.

3 R. Iraji, M. T. Manzuri-Shalmani, A. H. Jamalian, and H. Beigy, "IJA automaton: Expediency and ε-optimality properties," in *5th IEEE International Conference on Cognitive Informatics*, Beijing, China, pp. 617–622, 2006.

4 A. H. Jamalian, R. Iraji, A. R. Sefidpour, and M. T. Manzuri-Shalmani, "Examining the ε-optimality property of a tunable FSSA," in *6th IEEE International Conference on Cognitive Informatics*, Lake Tahoe, CA, USA, pp. 169–177, 2007.

5 K. S. Narendra and M. A. L. Thathachar, "Learning automata: An introduction," Englewood Cliffs, NJ: Prentice-Hall, 1989.

6 V. I. Varshavskii and I. P. Vorontsova, "On the behavior of stochastic automata with variable structure," *Automatika i Telemekhanika (USSR)*, vol. 24, pp. 327–333, 1963.

7 G. McMurtry and K. Fu, "A variable structure automaton used as a multi-modal searching technique," *IEEE Transactions on Automatic Control*, vol. 11, no. 3, pp. 379–387, 1966.

8 K. Fu and Z. Nikolic, "On some reinforcement techniques and their relation to the stochastic approximation," *IEEE Transactions on Automatic Control*, vol. 11, no. 4, pp. 756–758, 1966.

9 K. Fu, "11 stochastic automata as models of learning systems," *Mathematics in Science and Engineering*, vol. 66, pp. 393–431, 1970.

10 S. Lakshmivarahan, "Learning algorithms theory and applications," New York: Springer-Verlag, 1981.

11 K. Najim and A. S. Poznyak, "Learning automata: Theory and applications," New York: Pergamon, 1994.

12 B. J. Oommen and E. R. Hansen, "The asymptotic optimality of discretized linear reward-inaction learning automata," *IEEE Transactions on Systems, Man, and Cybernetics*, vol. 14, pp. 542–545, May/June 1984.

13 B. J. Oommen, "Absorbing and ergodic discretized two-action learning automata," *IEEE Transactions on Systems, Man, and Cybernetics*, vol. 16, pp. 282–296, May/June 1986.

14 M. A. L. Thathachar and P. S. Sastry, "A class of rapidly converging algorithms for learning automata," in *Proceedings of the IEEE International Conference on Cybernetics and Society*, Bombay, India, 1984, pp. 602–606.

15 M. A. L. Thathachar and P. S. Sastry, "A new approach to the design of reinforcement schemes for learning automata," *IEEE Transactions on Systems, Man, and Cybernetics*, vol. SMC-15, pp. 168–175, Jan/Feb 1985.

16 M. A. L. Thathachar and P. S. Sastry, "Estimator algorithms for learning automata," in *Proceedings of the Platinum Jubilee Conference on System and Signal Processing*, vol. Bangalore, India: Department of Electrical Engineering, December 1986, p. Indian Institute of Science.

17 J. K. Lanctôt and B. J. Oommen, "Discretized estimator learning automata," *IEEE Transactions on Systems, Man, and Cybernetics*, vol. 22, pp. 1473–1483, Nov/Dec 1992.

18 B. J. Oommen and J. K. Lanctot, "Discretized pursuit learning automata," *IEEE Transactions on Systems, Man, and Cybernetics*, vol. 20, pp. 931–938, July/Aug 1990.

19 M. S. Georgios I. Papadimitriou and A. S. Pomportsis, "A new class of ε-optimal learning automata," *IEEE Transactions on Systems, Man, and Cybernetics*, vol. 34, no. 1, pp. 246–254, 2004.

20 J. Zhang, Y. Wang, C. Wang, and M. Zhou, "Fast variable structure stochastic automaton for discovering and tracking spatiotemporal event patterns," *IEEE Transactions on Cybernetics*, vol. 48, no. 3, pp. 890–903, 2018.

21 Z. Zhang, D. Wang, and J. Gao, "Learning automata-based multiagent reinforcement learning for optimization of cooperative tasks," *IEEE Transactions on Neural Networks and Learning Systems*, vol. 32, no. 10, pp. 4639–4652, 2021.

22 S. Sahoo, B. Sahoo, and A. K. Turuk, "A learning automata-based scheduling for deadline sensitive task in the cloud," *IEEE Transactions on Services Computing*, vol. 14, no. 6, pp. 1662–1674, 2021.

23 A. Yazidi, I. Hassan, H. L. Hammer, and B. J. Oommen, "Achieving fair load balancing by invoking a learning automata-based two-time-scale separation paradigm," *IEEE Transactions on Neural Networks and Learning Systems*, vol. 32, no. 8, pp. 3444–3457, 2021.

24 X. Zhang, L. Jiao, O.-C. Granmo, and M. Goodwin, "On the convergence of Tsetlin Machines for the IDENTITY-and NOT operators," *IEEE Transactions on Pattern Analysis and Machine Intelligence*, vol. 44, no. 10, pp. 6345–6359, 2022.

3

Fast Learning Automata

An update scheme of a state probability vector of actions is critical for a learning automaton (LA). The most popular one is the pursuit scheme that pursues the estimated optimal action and penalizes others. The scheme was continuously improved in the past decades to accelerate the convergence of LAs. The complexity of the update scheme of a state probability vector is also critical to LAs' performance because it increases with the number of actions. This chapter introduces two LAs to accelerate the convergence and computational update of LAs. The first one achieves significantly faster convergence and higher accuracy than the classical pursuit scheme. The other lowers the computational complexity of updating a state probability vector to be independent of the number of actions. Together, they provide a fast learning method for the large-scale online reinforcement learning tasks of intelligent systems [1–3].

3.1 Last-position Elimination-based Learning Automata

This section first introduces a reverse philosophy opposed to the traditional pursuit scheme and leads to Last-position Elimination-based Learning Automata (LELA) [4] where the action graded last in terms of the estimated performance is penalized by decreasing its state probability and is eliminated when its state probability is decreased to zero. All active actions, i.e., actions with non-zero state probability, equally share the penalized state probability from the last-position action at each iteration. LELA is characterized by (i) "relaxed" convergence condition for its optimal action, (ii) "accelerated" step size of its state probability update scheme for the estimated optimal action, and (iii) "enriched" sampling for its estimated non-optimal actions. Last-position elimination is a widespread philosophy in the real world and is proved to be also helpful for the update

Learning Automata and Their Applications to Intelligent Systems, First Edition.
JunQi Zhang and MengChu Zhou.
© 2024 The Institute of Electrical and Electronics Engineers, Inc. Published 2024 by John Wiley & Sons, Inc.

scheme of LAs via the simulations of well-known benchmark environments. In the simulation studies, two versions of LELA using different selection strategies of the last action are compared with the classical pursuit schemes. Simulation results show that LELA achieves significantly faster convergence and higher accuracy than the classical pursuit schemes. Specifically, LELA reduces the interval to find the best parameter for a specific environment. Thus, its parameter tuning is easy to perform and its convergence time is much smaller than the classical ones when applied to a practical case. Their convergence curves and the corresponding variance coefficient curves are illustrated to characterize their essential differences and verify the analysis results of the introduced algorithms.

3.1.1 Background and Motivation

In LA designs, Variable Structure Stochastic Automata (VSSA) are introduced in [5] and characterized by the dynamical update of the state transition probabilities, while Fixed Structure Stochastic Automata (FSSA) [6, 7] use time-invariant transition and output functions. To speed up the convergence of VSSA, two approaches named as discretization and estimator are widely used. The discretization of a state probability space of actions is introduced in [8] by restricting the state probability of choosing an action to a finite number of values in the interval [0, 1] compared with the continuous update scheme of a probability vector. The estimator algorithms [9] use the running estimates of reward probabilities of actions rather than the feedback from an environment to change the state probability vector of choosing the next action. In nonestimator algorithms, the chosen action's probability increases if it is rewarded. However, estimator algorithms firstly update the estimate of the reward probability for the chosen action. The state probability vector is then updated by combining the estimated reward probability and the feedback from an environment. Based on the estimator, different update schemes for a state probability vector are proposed. The pursuit scheme [10] is the most popular one, which increases the state probability of the estimated optimal action but decreases others. The discretized pursuit scheme DP_{RI} is proposed in [11, 12] to speed up the continuous one [10]. The other famous one is the generalized pursuit scheme DGPA [13], which pursues the actions that have higher reward estimates than the current chosen action instead of a single estimated optimal action. The approaches presented in [10–13] have substantially improved the rate of LAs convergence. Based on the pursuit scheme, a stochastic estimator (SE) is used to further speed up LAs convergence by utilizing a Reward–Inaction paradigm (SE_{RI}) in a stationary and random environment [14].

Last-position Elimination-based Learning Automaton (LELA) is introduced in this section. An automaton interacts with an environment. The latter consists of three components denoted by $\langle A, B, D \rangle$. An estimator-based LA is described by a quadruple $\langle A, B, Q, L \rangle$, where

1) $A = \{\alpha_1, \dots, \alpha_r\}$ is the set of actions ($2 \le r < \infty$). The action selected at time t is denoted by $\alpha(t)$.
2) $B = \{0, 1\}$ is the input set of possible environmental responses. "1" and "0" denote the reward and penalty responses, respectively. The environmental response at time t is denoted by $\beta(t)$.
3) $D = (d_1, \dots, d_r)$ where d_i ($i \in \mathbb{Z}_r = \{1, 2, \dots, r\}$) is the reward probability vector that the environment rewards actions. The environment is assumed to be stationary. Hence D is constant for all the time. The optimal action with the maximum probability $\max\{d_i\}$, ($i \in \mathbb{Z}_r$) is unique as the classical constraint.
4) Q is the vector state of the automaton. Furthermore, the state consists of two r-tuples namely P and $\tilde{D}(t)$. When Q or any of its components is indexed by t, it represents the vector state of the automaton at time t. P is a probability distribution over the set of actions. We have $P(t) = (p_1(t), \dots, p_r(t))$, where $p_i(t)$ is the probability of selecting action α_i at t and it satisfies $\sum_{i=1}^{r} p_i(t) = 1$. $\tilde{D}(t) = (\tilde{d}_1(t), \dots, \tilde{d}_r(t))$ is the estimator vector that contains the current estimates of the reward probabilities of actions at t.
5) L is the learning scheme to update Q.

In order to avoid confusion, we refer above probability notations to the following names.

$p_i(t)$: state probability of action i at time t
d_i: reward probability of action i
$\tilde{d}_i(t)$: estimated reward probability of action i at time t

We compute $\tilde{d}_i(t)$ as follows:

$$\tilde{d}_i(t) = \frac{\mathbb{H}_i(t)}{\mathbb{G}_i(t)}. \tag{3.1}$$

where $\mathbb{G}_i(t)$ is the number of times that action α_i has been selected up to time instant t, and $\mathbb{H}_i(t)$ is the sum of the environmental responses received up to time t.

In the classical pursuit algorithm DP_{RI} [11, 12], when an action is rewarded, all the actions that do not correspond to the highest estimate have their state probabilities decreased by a step $\Delta = \frac{1}{m}$, where r is the number of allowable actions and n is a resolution parameter. In order to keep the sum of the components of the vector $P(t)$ equal to unity, the state probability of the action with the highest estimate has to be increased by an integral multiple of the smallest step size Δ. When the action chosen is penalized, there is no update in the state probability vector. DPRI is thus under the *reward–inaction* paradigm. Assume that m is the index of the estimated optimal action. From [11, 12], we have the following update scheme of DP_{RI}:

If $\beta(t) = 1$ Then

$$p_j(t+1) = \max_{j \neq m} \{p_j(t) - \Delta, 0\}, \forall j \in \mathbb{Z}_r. \tag{3.2}$$

$$p_m(t+1) = 1 - \Sigma_{j \neq m} p_j(t+1). \tag{3.3}$$

Else

$$p_j(t+1) = p_j(t), \forall j \in \mathbb{Z}_r. \tag{3.4}$$

Endif

Example 3.1 In DP_{RI}, there are five actions or $r = 5$. $A = \{\alpha_1, \ldots, \alpha_5\}$. At $t = 0$, their reward probability vector $D = \{0.85, 0.8, 0.75, 0.6, 0.5\}$, their probability distribution $P(0) = \{0.2, 0.2, 0.2, 0.2, 0.2\}$, after 10 times of updates, their estimated reward probability vector $\tilde{D}(0) = \{0.81, 0.82, 0.74, 0.65, 0.48\}$. Let $n = 2$, $\Delta = \frac{1}{m} = \frac{1}{5 \times 2} = 0.1$. If at $t = 1$, α_3 is chosen and its reward $\beta(1) = 1$, the estimated reward probability $\tilde{d}_3(1)$ is updated and let it be 0.78. The current estimated optimal action α_2 is rewarded and other actions are penalized. Therefore, their probability distribution $P(1) = (p_1(1) = 0.1, p_2(1) = 0.6, p_3(1) = 0.1, p_4(1) = 0.1, p_5(1) = 0.1)$. This example shows that the pursuit scheme of DP_{RI} pursues the current estimated optimal action and penalizes others.

The generalized pursuit algorithm DGPA [13] pursues the actions that have higher reward estimates than the currently chosen action instead of a single estimated optimal action. Its update scheme is as follows [13].

$$p_j(t+1) = \min \left\{ p_j(t) + \frac{\Delta}{\mathbb{K}(t)}, 1 \right\}, \forall j \in \mathbb{Z}_r, j \neq i \tag{3.5}$$
$$\text{such that } \tilde{d}_j(t) > \tilde{d}_i(t)$$

$$p_j(t+1) = \max \left\{ p_j(t) - \frac{\Delta}{r - \mathbb{K}(t)}, 0 \right\}, \forall j \in \mathbb{Z}_r, j \neq i \tag{3.6}$$
$$\text{such that } \tilde{d}_j(t) < \tilde{d}_i(t)$$

$$p_i(t+1) = 1 - \sum_{j \neq i} p_j(t+1), \forall j \in \mathbb{Z}_r, \tag{3.7}$$

where $\mathbb{K}(t)$ denotes the number of actions that have higher estimates than the chosen action at time t and i is the index of the current chosen action.

Example 3.2 In DGPA, there are five actions or $r = 5$. $A = \{\alpha_1, \dots, \alpha_5\}$. At $t = 0$, their reward probability vector $D = \{0.85, 0.8, 0.75, 0.6, 0.5\}$, their probability distribution $P(0) = \{0.2, 0.2, 0.2, 0.2, 0.2\}$, after 10 times of updates, their estimated reward probability vector $\tilde{D}(0) = \{0.81, 0.82, 0.74, 0.65, 0.48\}$. Let $n = 2$, and $\Delta = \frac{1}{rn} = \frac{1}{5 \times 2} = 0.1$. If at $t = 1$, α_3 is chosen and its reward $\beta(1) = 1$, the estimated reward probability $\tilde{d}_3(1)$ is updated and let it be 0.78. The actions whose estimated reward probability is greater than $\tilde{d}_3(1)$, i.e., α_1 and α_2, are rewarded and other actions are penalized. Therefore, their probability distribution, i.e., $P(1) = \{0.25, 0.25, 0.2, 0.15, 0.15\}$. This example shows that the pursuit scheme of DGPA pursues the actions that have higher reward estimates than the currently chosen action instead of a single estimated optimal action.

DP$_{RI}$ and DGPA have been proved to be ϵ-optimal in every stationary random environment [11–13].

3.1.2 Principles and Algorithm Design

LELA uses the following update scheme:

If $\beta(t) = 1$ Then

    Find $l \in \mathbb{Z}_r$ such that

$$\tilde{d}_l(t) = \min\{\tilde{d}_i(t) | p_i(t) \neq 0\}, \ i \in \mathbb{Z}_r \tag{3.8}$$

$$p_l(t+1) = \max\{p_l(t) - \Delta, 0\} \tag{3.9}$$

    If $p_l(t+1) = 0$ Then $\tilde{r} = \tilde{r} - 1$ Endif $\tag{3.10}$

$$p_j(t+1) = \min\left\{ p_j(t) + \frac{p_l(t) - p_l(t+1)}{\tilde{r}}, 1 \right\},$$

$$\forall j \in \mathbb{Z}_r, \text{ such that } p_j(t) > 0. \tag{3.11}$$

Endif

Note that the active actions are the actions with positive state probability $p_i(t) > 0$, $i \in \mathbb{Z}_r$. $\tilde{r}$ denotes the number of the active actions and is initialized as r. l is the index of the last-position active action with the minimal estimated reward probability. It is notable that only the active actions' state probabilities are updated. If there are several active actions with identical minimal estimated reward probabilities, we employ two strategies to choose one from them as the last-position action. The first strategy returns the one with the smallest action index, while the second one chooses it randomly. The performances of these two strategies are presented in the simulation section.

It is worth noting that the number of allowable actions r and the resolution parameter n are constant in LELA such that the step size of its learning step size $\Delta = \frac{1}{m}$ is constant. When the last action α_l is eliminated, its $p_l(t)$ becomes 0 and the number of the active actions $\tilde{r}$ decreases by 1 while Δ does not change. The decreased $\tilde{r}$ increases the shared state probability of the active actions from the current last-position action. The algorithm of LELA is as follows.

LELA: Last-position Elimination-based Learning Automata Algorithm
Parameters

l Index of the minimal component of the reward estimate vector $\tilde{D}(t)$,
$l = \min_i \{ \tilde{d}_i(t) | p_i(t) \neq 0, i \in \mathbb{Z}_r \}$;

$\mathbb{H}_i(t)$ Number of times the ith action has been rewarded up to time t, $\forall i \in \mathbb{Z}_r$;

$\mathbb{G}_i(t)$ Number of times the ith action has been chosen up to time t, $\forall i \in \mathbb{Z}_r$;

n Resolution parameter;

$\Delta = \frac{1}{m}$ Smallest step size.

Method
Input: Number of allowable actions r, resolution parameter n, action set A, environmental response set B.
Output: Estimated optimal action α_m.
Initialize $p_i(0) = 1/r, \forall i \in \mathbb{Z}_r$
Initialize $\tilde{d}_i(t)$ by choosing each action for a number of times.
1: At time t pick $\alpha(t)$ according to probability distribution $P(t)$. Let $\alpha(t) = \alpha_i$.
2: Receive feedback $\beta_i(t) \in \{0, 1\}$. Update $\tilde{d}_i(t)$ according to the following equations for the action chosen

$$\mathbb{H}_i(t+1) = \mathbb{H}_i(t) + \beta_i(t)$$
$$\mathbb{G}_i(t+1) = \mathbb{G}_i(t) + 1$$
$$\tilde{d}_i(t+1) = \mathbb{H}_i(t+1)/\mathbb{G}_i(t+1)$$

3: If $\beta_i(t) = 0$ **Then**
    $p_i(t+1) = p_i(t) \; \forall j \in \mathbb{Z}_r$
    Goto Step 1.
    Endif
4: Update $P(t)$ according to the update schemeof LELA in (3.8) ~ (3.11)
5: If $\max\{P(t)\} = 1$, **Then** CONVERGE to the action α_m whose $p_m = \max\{P(t)\}$. **Else** *Goto* step 1. **Endif**
END

Example 3.3 In LELA, there are five actions, i.e., $r = 5$, and action set $A = \{\alpha_1, \ldots, \alpha_5\}$. At $t = 0$, their reward probability vector $D = \{0.85, 0.8, 0.75, 0.6, 0.5\}$, their probability distribution, i.e., $P(0) = \{0.2, 0.2, 0.2, 0.2, 0.2\}$, After 10 times of updates, their estimated reward probability vector $\tilde{D}(0) = \{0.81, 0.82, 0.74, 0.65, 0.48\}$. Let $n = 2$, $\Delta = \frac{1}{m} = \frac{1}{5 \times 2} = 0.1$. If at $t = 1$, α_3 is chosen and its reward $\beta(1) = 1$, the estimated reward probability $\tilde{d}_3(1)$ is updated and let it be 0.78. The actions whose estimated reward probability is greater than 0 but is the last one in the active actions are penalized. Therefore, their probability distribution, i.e., $P(1) = \{0.25, 0.25, 0.25, 0.25, 0.1)\}$. This example shows that the pursuit scheme of LELA penalizes the last-position action.

3.1.3 Difference Analysis

The reinforcement signal of LAs is probabilistic and noisy examples. The error as the probability of not choosing the optimal action and the convergence speed can be jointly affected by the convergence condition for the optimal action, step size of the state probability update for the action with the maximal estimated reward probability, and sampling strategy of actions. The convergence condition for the optimal action determines the condition under which the optimal action can be pursued. The step size directly affects the convergence speed. The error as the probability of not choosing the optimal action is mainly led by the insufficient sampling of some actions. These three aspects correlate to each other and jointly affect the convergence speed and accuracy. Wrong pursuit not only reduces the accuracy, but also lowers the convergence speed of LA because wrong decrease of the optimal action's probability needs additional iterations to be compensated for the right convergence. The essential differences between LELA and the classical pursuit schemes DP_{RI} and DGPA are identified in the following three aspects.

Assume m is the index of the optimal action. In the classical pursuit schemes, the convergence condition for the optimal action is that $\tilde{d}_m(t) > \tilde{d}_i(t)$ for all i such that $i \neq m$ and all $t \geq t_0$. For DP_{RI}, all actions except the one with the maximal estimated reward probability have to be penalized. For DGPA, $\tilde{d}_m(t)$ also has to be the maximal estimated reward probability among the ones of "all" actions because any action $i \neq m$ can be chosen such that $\tilde{d}_m(t)$ has to be greater than all $\tilde{d}_i(t)$ for convergence.

While in LELA, as long as $\tilde{d}_m(t) > \tilde{d}_l(t)$ where l is the index of the active action with the minimal estimated reward probability, the state probability $p_m(t)$ of the optimal action can be increased. Hence, LELA relaxes the convergence condition of the optimal action because only the active action graded last in the estimated performance has its state probability decreased.

We investigate the differences of the step size of the update scheme for $p_m(t)$ between LELA and the classical pursuit schemes, i.e., DP_{RI} and DGPA. Assume that m is the index of the action with the largest estimated reward probability $\tilde{d}_m(t)$.

DP_{RI} always pursues the estimated optimal action and $p_m(t)$ is updated as (3.2) $\sim$ (3.4). Hence, the step size of $p_m(t)$ is

$$\hat{\Delta}_{DP_{RI}} = \sum_{i=1,i\neq m}^{r} \min\{\Delta, \, p_i(t)\}. \tag{3.12}$$

This equation indicates that the step size of the update scheme for $p_m(t)$ in DP_{RI} "decreases" because more and more $p_i(t)$ turns to be zero as the LA evolves.

DGPA pursues the actions that have higher reward estimates than the currently chosen action instead of a single estimated optimal action. $p_m(t)$ is updated according to (3.5) $\sim$ (3.7). The step size of $p_m(t)$ can be written as

$$\hat{\Delta}_{DGPA} = \begin{cases} \sum_{i=1,i\neq m}^{r} \min\left\{p_i(t), \dfrac{\Delta}{r}\right\}, & \text{if } \mathbb{K}(t) = 0; \\ \dfrac{\Delta}{\mathbb{K}(t)}, & \text{if } \mathbb{K}(t) \geq 1 \end{cases} \tag{3.13}$$

where $\mathbb{K}(t)$ denotes the number of actions that have higher estimates than the chosen action at time t and i is the index of the current chosen action. When the m-th action is chosen, $\mathbb{K}(t) = 0$ such that each action $i \neq m$ is penalized by a constant $\frac{\Delta}{r}$ as same as the way in DP_{RI}. Otherwise, the probability $p_m(t)$ is increased by $\frac{\Delta}{\mathbb{K}(t)}$.

Equation (3.13) clarifies that the step size of the update scheme for $p_m(t)$ in DGPA consists of two cases. The first case is that the m-th action is chosen such that the step size of $p_m(t)$ adopts the same "decreasing" step size of DP_{RI}. All actions except the m-th action decrease their action probabilities by a constant step size $\frac{\Delta}{r}$. Only the m-th action with the largest estimated reward probability $\tilde{d}_m(t)$ is pursued. The second case is that the m-th action is not chosen such that the actions that have higher reward estimates than the currently chosen action are pursued evenly. With the number of actions that have higher estimates than the chosen action decreasing, $\mathbb{K}(t)$ decreases such that step size "increases."

The second case of the update scheme for $p_m(t)$ in DGPA seems similar to the update scheme for $p_m(t)$ in LELA. However, the condition for an action to be pursued changes in essence. In LELA, the action's probability can be increased if only its estimated reward probability is higher than that of the action with the least positive probability.

Thus, the step size of $p_m(t)$ in DGPA can be written as

$$\hat{\Delta}_{DGPA} = p_m(t) \sum_{i=1,i\neq m}^{r} \min\left\{p_i(t), \frac{\Delta}{r}\right\} + (1 - p_m(t))\frac{\Delta}{\mathbb{K}(t)}. \tag{3.14}$$

In its early stage of iterations, these two cases are mixed to update the action probabilities. In its middle and late periods of the learning process, DGPA mainly pursues the estimated optimal action because it has larger and larger state probability $p_m(t)$ to be sampled. Hence, the philosophy of DGPA in the middle and late learning periods tends to adopt the same "decreasing" step size of DP_{RI}.

In LELA, only the action graded last in the estimated performance has its state probability decreased and finally is laid off from its position. All active actions except the last one are all equally pursued to share the state probability from the last-position action at each iteration. Hence,

$$\hat{\Delta}_{LELA} = \frac{p_l(t) - p_l(t+1)}{\tilde{r}} = \frac{\min\{\Delta, \ p_l(t)\}}{\tilde{r}}. \tag{3.15}$$

Above equation indicates that the step size of the update scheme for $p_m(t)$ in LELA "increases" because more and more $p_i(t)$ turns to be zero such that the number of the active actions $\tilde{r}$ decreases with the learning process. It means that the step size in LELA is "accelerated" along with the learning process. It is reasonable because the number of times that active actions are selected increases and the reward probability estimation of active actions becomes more and more reliable along with the learning process. Such "accelerated" step size of the update scheme for $p_m(t)$ in LELA is essentially different from those in DP_{RI} and DGPA.

The above equation of LELA also indicates that, from the early period to the late one, "all" active actions equally share the decreased state probability from the last active action, which is different from the two-case mixed pursuit scheme in DGPA and the optimal-oriented pursuit scheme in DP_{RI}. This strategy avoids the excessively repeated sampling of the estimated optimal action such that other actions have more chances to show their competencies. Consequently, the real optimal action can be more easily learned out of such more equitable competition.

Table 3.1 Accuracy (number of correct convergences/number of experiments).

	d_1	d_2	d_3	d_4	d_5	d_6	d_7	d_8	d_9	d_{10}
E_1	**0.65**	0.50	0.45	0.40	0.35	0.30	0.25	0.20	0.15	0.10
E_2	**0.60**	0.50	0.45	0.40	0.35	0.30	0.25	0.20	0.15	0.10
E_3	**0.55**	0.50	0.45	0.40	0.35	0.30	0.25	0.20	0.15	0.10
E_4	**0.70**	0.50	0.30	0.20	0.40	0.50	0.40	0.30	0.50	0.20
E_5	0.10	0.45	**0.84**	0.76	0.20	0.40	0.60	0.60	0.50	0.30

Bold values are the max reward probability.

3.1.4 Simulation Studies

This section first compares the relative performances of the LELA with the classical pursuit algorithms DP_{RI} and DGPA by presenting their accuracy and convergence speed. To be fair, the stochastic estimator, based SE_{RI} [14] is not compared because it needs one more parameter for its stochastic estimator, which is sensitive and arduous to be found for each environment. We, hence, focus on comparing the difference between the pursuit scheme and the introduced last-position elimination scheme.

$LELA_S$ and $LELA_R$ are LELA using two strategies, respectively, to choose one from the active actions with identical minimal estimated reward probability as the last-position action. $LELA_S$ returns the one with the smallest action index, while $LELA_R$ chooses it randomly. The convergence curves and the corresponding variance coefficient curves of the contenders are then illustrated to characterize their essential differences and verify the analysis results in Section 3.1.3.

The popular benchmark test sets with ten actions are used to examine the performance of LELA. E_1–E_3 are popular benchmarks. E_4 and E_5 are also the well-known benchmark environments and were used in [11–15]. The actions' probabilities for each environment are presented in Table 3.1. Each environment includes ten allowable actions. A LA aims at learning the optimal action from the environment with the maximum reward probability by its interaction with a random environment. During a cycle, it chooses an action and then receives a stochastic response that can be either a reward ($\beta_i(t) = 1$) or penalty ($\beta_i(t) = 0$) from the environment. It employs this response and updates the state probability vector of choosing the next action.

An algorithm is considered to have converged if the probability of choosing an action is not less than a threshold $\hat{T}$ ($0 < \hat{T} \leq 1$) in all the performed tests. If the automaton converges to the action that has the highest reward probability, it is considered to have converged correctly.

As a common sense, the step size is a trade-off between the convergent accuracy and speed. Before comparing their performances, a large number of tests using the same method in [8, 11–13] are executed to determine the "best" values of the resolution parameter n for each scheme. The values are considered as the best values if they yield the fastest convergence and the automaton converges to the correct action in a sequence of N_E experiments. These best parameters are then chosen as the final parameter values used for the respective algorithms to compare their performances. The values of $\hat{T}$ and N_E are taken to be equal to the ones used in [8, 11–13]. Thus, $\hat{T} = 0.999$ and $N_E = 750$.

Each result is obtained from 250 000 runs for each algorithm in each environment by using the best learning parameters. Every algorithm samples all actions ten times each in order to initialize the estimated reward probability vector. These

Table 3.2 Accuracy (number of correct convergences/number of experiments).

	LELA$_S$	LELA$_R$	DGPA	DP$_{RI}$
E_1	0.997	0.997	0.997	0.995
E_2	0.996	0.996	0.996	0.994
E_3	0.995	0.995	0.995	0.993
E_4	0.997	0.997	0.997	0.996
E_5	0.997	0.997	0.997	0.994

extra iterations are also included in the results presented in the following tables and figures.

Accuracy is defined as the ratio between the number of correctly converged ones and the total number of experiments. Convergence speed is measured by the average number of iterations required for convergence.

Table 3.2 presents the accuracy of the contenders. Both LELA$_S$ and LELA$_R$ achieve higher accuracy than the classical pursuit algorithms DP$_{RI}$ and DGPA even though the difference is insignificant.

The convergence speed of the contenders is presented in Table 3.3. It is shown that the LELA$_S$ and LELA$_R$ have faster convergence speed than both DP$_{RI}$ and DGPA in all of the benchmark environments. In Table 3.3, "Para" denotes the best parameter of contenders, and "Ite" denotes the convergence speed which is the mean iterations needed for convergence.

The speedup rate of LELA$_S$ compared with the classical pursuit scheme-based DP$_{RI}$ ($(\mathrm{Ite}_{DP_{RI}}-\mathrm{Ite}_{LELA_S})/\mathrm{Ite}_{DP_{RI}}$) is {42.09%, 50.08%, 65.94%, 25.82%, 52.75%} in

Table 3.3 The average number of iterations required for convergence('Para' denotes the best parameter of contenders and 'Ite' denotes the convergence speed).

	LELA$_S$		LELA$_R$		DGPA		DP$_{RI}$	
	Para	Ite	Para	Ite	Para	Ite	Para	Ite
E_1	$n = 9$	630	$n = 9$	629	$n = 33$	880	$n = 298$	1088
E_2	$n = 19$	1243	$n = 17$	1129	$n = 65$	1677	$n = 653$	2490
E_3	$n = 60$	3791	$n = 59$	3733	$n = 204$	5187	$n = 3221$	11132
E_4	$n = 9$	586	$n = 9$	586	$n = 28$	755	$n = 216$	790
E_5	$n = 25$	1114	$n = 24$	1072	$n = 55$	1446	$n = 881$	2358

benchmarks from $E_1 - E_5$. Compared with the generalized pursuit scheme-based DGPA, LELA$_S$ achieves the speedup rate {28.41%, 25.88%, 26.91%, 22.38%, 22.96%}. These are very significant.

More significant results are obtained with LELA$_R$. Its speedup rate compared with the classical pursuit scheme-based DP$_{RI}$ is {42.19%, 54.66%, 66.47%, 25.82%, 54.54%} in benchmarks from $E_1 - E_5$. Compared with the generalized pursuit scheme-based DGPA, it obtains the speedup rate {28.523%, 32.68%, 28.03%, 22.38%, 25.86%}.

Therefore, both LELA$_S$ and LELA$_R$ converge faster than the classical pursuit algorithms DP$_{RI}$ and DGPA.

Apart from the better performance of LELA$_S$ and LELA$_R$, it is worth noticing that LELA$_S$ and LELA$_R$ decrease the interval to find the best parameter for a specific environment in the classical pursuit algorithms DP$_{RI}$ and DGPA. In DGPA, the best parameter ranges from 28 to 204 in five environments. In DP$_{RI}$, the best parameter ranges from 216 to 2356. By contrast, in LELA$_S$ and LELA$_R$, the best one ranges from 9 to 60 only. Specially, in E_1 and E_4, the best parameters are both 9. This interesting result shows that LELA$_S$ and LELA$_R$ can have their parameter tuning easier to perform and can save much more time when applied to a practical case than the classical ones.

In addition to the numerical results, the convergence curves and corresponding variance coefficient curves of the LELA$_R$ and the classical pursuit algorithms DP$_{RI}$ and DGPA are illustrated in Figs. 3.1 and 3.2, respectively, to characterize their essential differences and verify the analysis results in the above section. The convergence curve represents the mean choice probability $p_m(t)$ of the optimal action as a function of the number of iterations. The higher $p_m(t)$, the faster convergence to the optimal action because the LA algorithm is considered to have converged if the probability of choosing an action is not less than a threshold $\hat{T}$ ($0 < \hat{T} \leq 1$) in all the performed tests.

The variance coefficient curve depicts the variance coefficient [14] (the variance coefficient of a random variable $\tilde{X}$ is defined as the fraction $\sigma[\tilde{X}]/E[\tilde{X}]$, where $\sigma[\tilde{X}]$ is the standard deviation, and $E[\tilde{X}]$ is the mean value of random variable $\tilde{X}$) of the variable $p_m(t)$ along with the learning process. The variance coefficient expresses the standard deviation of the random variable as a percentage of its mean value. Bigger variance coefficient means the $p_m(t)$ varies more in multiple runs at the t-th iteration. When $E[p_m(t)] = \hat{T}$, bigger variance coefficient means the convergence speeds vary more in multiple runs.

Compared with the classical pursuit algorithms DP$_{RI}$ and DGPA, Fig. 3.1 shows that the LELA$_R$ has not only faster convergence speed, but also an "accelerated" step size of the update scheme for $p_m(t)$. Its increase of the $p_m(t)$ is accelerated along with the learning process such that the speed increase of $p_m(t)$ is slower in

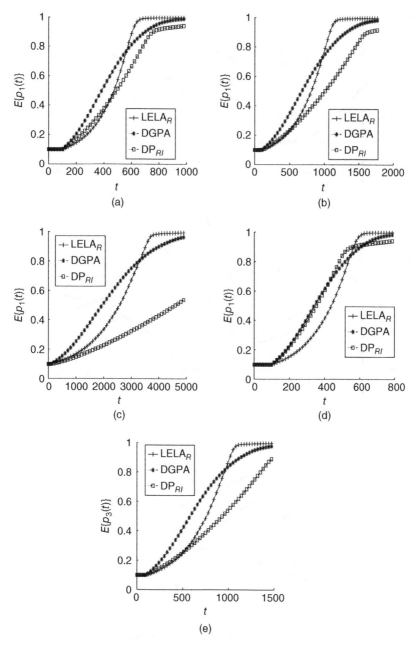

Figure 3.1 Mean of the Action probability $E\{P_m(t)\}$ of the optimal action versus t. (a) E_1; (b) E_2; (c) E_3; (d) E_4; (e) E_5.

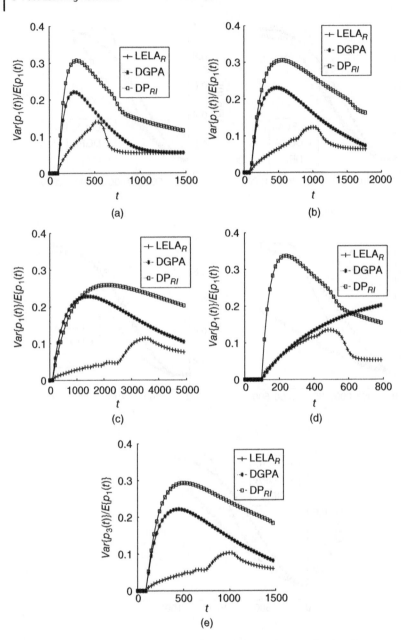

Figure 3.2 Variance coefficiency $Var\{P_m(t)\}/E\{P_m(t)\}$ of the optimal action versus t. (a) E_1; (b) E_2; (c) E_3; (d) E_4; (e) E_5.

the early stage and faster in the late stage of the convergence than the ones in classical pursuit schemes. Such faster convergence speed and accelerated convergence curves verify that the $LELA_R$ has an accelerated step size of the update scheme for $p_m(t)$ compared with the classical pursuit algorithms. Although the LELA converges faster only in the late stage when $\hat{T} > 0.8$, it satisfies the real needs because $\hat{T}$ is generally set to be not less than 0.9 [11–14].

Figure 3.2 shows that $LELA_R$ has lower variance coefficient curves than the classical pursuit algorithms along with the learning process in all test environments, which satisfies the intense need for the reliable estimates and leads to a higher performance improvement. The lower variance coefficient curves of $LELA_R$ also verify the above analysis results that $LELA_R$ has the relaxed condition for the convergence to the optimal action and gives "all" of the non-optimal estimated active actions more sufficient sampling from the early period to the late one of the learning process such that the competition among actions becomes more equitable and the variance coefficient of $p_m(t)$ becomes smaller. This strategy avoids the excessively repeated sampling of the estimated optimal action such that other actions have more chances to show their competencies. Furthermore, the lower variance coefficient curve of $LELA_R$ maintains more stable convergence speed for each run in multiple runs.

3.1.5 Summary

The pursuit scheme has been an outstanding and mainstream update scheme for the estimator-based VSSA since a few decades ago. This section introduces a reverse philosophy to update the action probabilities, named as last-position elimination-based learning automaton (LELA). It has a looser and fair rule for the action competition to avoid the excessively repeated sampling of the estimated optimal action. This philosophy actually is widespread in the real world and is proved to be also helpful for the update scheme of LAs via the simulations of well-known benchmark environments. The proof of the ϵ-optimal property of LELA can refer [4]. Compared with the classical pursuit scheme, the last-position elimination scheme possesses three more competitive characteristics, i.e., the "relaxed" convergence condition for the optimal action, the "accelerated" step size of state probability update of action with the maximal estimated reward probability, and the "enriched" sampling for the estimated non-optimal actions. These three characteristics cooperatively contribute to the performance improvement in terms of accuracy and convergence speed by reducing the probability of false selection.

In the simulations, two versions of LELA using different selection strategies of the last action when one has several identical minimum values of estimated reward probability are compared with the classical pursuit algorithms.

The random selection strategy obtains a little faster convergence speed than the smallest action index selection strategy. Both of them achieve significantly faster convergence speed and higher accuracy than classical pursuit algorithms. Moreover, they reduce the interval to find the best parameter for a specific environment in classical pursuit algorithms. Hence, they can have their parameter tuning easier to perform and can save much more time when applied to a practical case than classical ones. In addition to the numerical results, the convergence curves and the corresponding variance coefficient curves of the contenders are illustrated to characterize their essential differences and verify the analysis results. The last-position elimination scheme can also incorporate into itself other methods like the stochastic estimator.

3.2 Fast Discretized Pursuit Learning Automata

The discretized pursuit LA is the most popular one among variants of Learning automata. During an iteration its operation consists of three basic phases: (1) selecting the next action, (2) finding the optimal estimated action, and (3) updating the state probability. However, when the number of actions is large, the learning becomes extremely slow because there are too many updates to be made at each iteration. The increased updates are mostly from phases 1 and 3. A fast discretized pursuit LA with assured ε-optimality [16] is introduced to perform both phases 1 and 3 with a computational complexity independent of the number of actions. Apart from its low computational complexity, it achieves faster convergence speed than the classical one when operating in stationary environments. This variant can promote the applications of learning automata toward the large-scaleaction-oriented area that requires efficient reinforcement tools with assured ε-optimality, fast convergence speed, and low computational complexity for each iteration.

3.2.1 Background and Motivation

With respect to their Markovian representations, learning automata are classified into two main categories: ergodic or automata-possessing absorbing barriers. The former converge with a distribution independent of the initial state. On the other hand, the latter converge to a particular action after a number of finite steps. If the reward probabilities of actions are stable (stationary environment), absorbing automata are preferred. The linear reward-inaction (denoted L_{RI}) scheme was originally reported to be ε-optimal in all stationary random environments. The linear reward–penalty scheme (denoted L_{RP}) is known expedient in stationary random environments and is more preferred for non-stationary environments.

In order to increase the speed of convergence, two approaches named as discretization and estimator are widely used. The discretization of the state probability space of actions is introduced in [8] by restricting the state probability of choosing an action to a finite number of values in the interval [0, 1] compared with the continuous update scheme of the probability vector. The estimator algorithms [9] use the running estimates of the reward probabilities of actions rather than the feedback from the environment to change the state probability vector of choosing the next action. In the nonestimator algorithms like L_{RI} and L_{RP}, the chosen action's probability increases if it is rewarded. However, the estimator algorithms firstly update the estimate of the reward probability for the chosen action. The state probability vector is then updated by combining the estimated reward probability and feedback from the environment. Based on the estimator, different update schemes for a state probability vector are introduced. The pursuit scheme [10] is the most popular one, which increases the state probability of the estimated optimal action but decreases others. The discretized pursuit scheme DP_{RI} is introduced in [11, 12] to speed up the continuous one [10]. At each iteration, its operation consists of: (i) selecting an action, (ii) finding the optimal estimated action, and (iii) updating the state probability.

However, for a large number of actions, their convergence becomes extremely slow because many action state probabilities must be updated every instant. This limitation prevents the learning automata from the large-scale-action oriented applications. In order to overcome the high dimensionality of the decision space, hierarchical systems of learning automata are introduced [17, 18]. In [17], several learning automata are combined in order to construct another one that has a much larger number of actions by joining them in a hierarchical mode and a tree structure. Every action of each subautomaton (except the last level ones) corresponds to a subautomaton of the next level of the tree. The actions of the last level interact with the environment directly. The reaction of the environment is used to update the action probabilities at all the levels of hierarchy by using the non-pursuit scheme L_{RI}. In [18], the pursuit scheme is used for the hierarchical learning automata to speed up the convergence. The computational complexity of the probability updating is $O(r)$ for DP_{RI} and is reduced to $O(\log r)$ in hierarchical automata where r is the number of actions.

A major source of motivation for the following presented fast discretized LAs is the Fast Learning Automata (FLA) [19], which is introduced for non-stationary environments with the capability of performing both probability updating and action selection fast, with a computational time independent of the number of actions. However, FLA is not ε-optimal and cannot be used in a stationary environment because FLA just updates the probability of the chosen action regardless of others. Moreover, it is based on L_{RP} scheme whose convergence speed is much slower than the pursuit scheme.

This section introduces a fast discretized pursuit LA with assured ε-optimality to perform both action selection and state probability update with a computational complexity independent of the number of actions. Apart from its low computational complexity, it achieves faster convergence speed than the classical one when operating in a stationary environment. This LA can promote the applications of learning automata toward the large-scale-action oriented area that requires efficient reinforcement tools with assured ε-optimality, fast convergence speed, and low or even real-time computational complexity for each iteration.

3.2.2 Algorithm Design of Fast Discretized Pursuit LAs

First of all, an example is given to illustrate the way of state probability update in DP_{RI}. It is exactly such illustration that inspires the fast discretized pursuit LAs.

Example 3.4 In the classical discretized pursuit algorithm DP_{RI} [11, 12], when an action is rewarded, all the actions that do not correspond to the highest estimate have their state probabilities decreased by a step $\Delta = \frac{1}{m}$ where r is the number of allowable actions and n is a resolution parameter. In order to keep the sum of the components of vector $P(t)$ equal to unity, the state probability of the action with the highest estimate has to be increased by an integral multiple of the smallest step size Δ. When the chosen action is penalized, there is no updating in the state probability vector, and it is thus of the reward–inaction paradigm. Once an update occurs, $(r - 1) \times \Delta$ are rewarded to the estimated optimal action and other actions are all penalized by Δ. Figure 3.3 illustrates this update scheme. Each Δ is considered as a Probability Cell (PC) shown in Fig. 3.3 (a), where each action is initialized to n cells, i.e., $n\Delta$. Assume that α_1 is the estimated optimal action and positive response is received, $(r - 1)$ probability cells are rewarded to α_1 and one probability cell is penalized for other actions as shown in Fig. 3.3 (b), respectively. The penalty for these actions can be illustrated as one of its PCs is rewritten by α_1.

The introduced Fast Discretized Pursuit learning automaton (FDP_{RI}) exploits the most significant pattern of the discretized update scheme in DP_{RI}. This pattern always increases the state probability of the estimated optimal action but decreases others by a discretized step Δ.

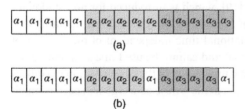

(a)

(b)

Figure 3.3 DP_{RI} using $r = 3$ and $n = 5$, α_1 is the estimated optimal action and positive response is received. (a) DP_{RI}: Initialization. (b) DP_{RI}: State probability update when the chosen action is rewarded.

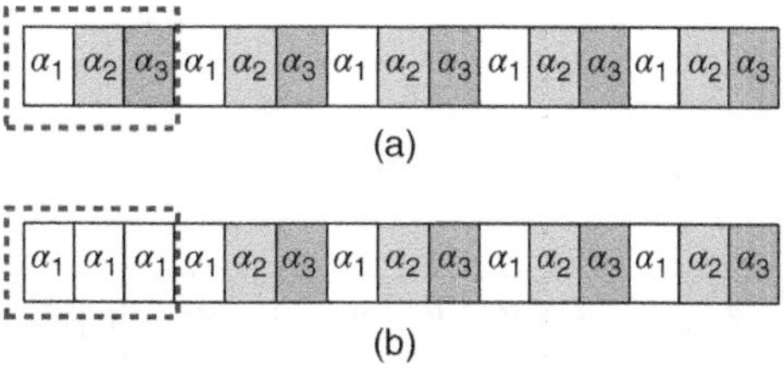

Figure 3.4 FDP_{RI} using $r = 3$ and $n = 5$, α_1 is the estimated optimal action and it received positive response. (a) FDP_{RI}: Initialization. (b) FDP_{RI}: Rewarded when $\mathbb{C} \geq 1$.

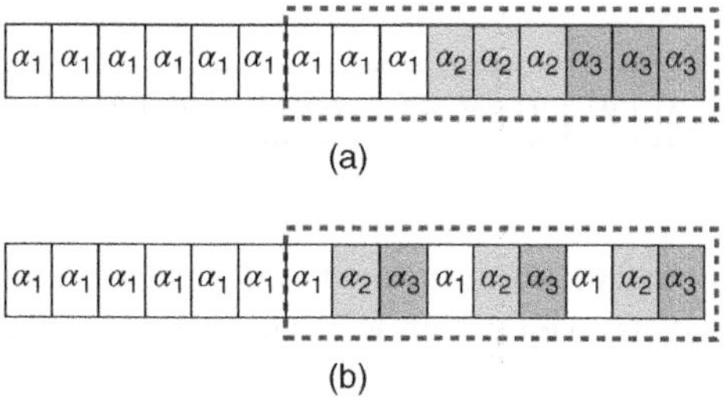

(a)

(b)

Figure 3.5 $\mathbb{U}$-$\mathbb{C}$ Transformation: $r = 3$ and $n = 5$. (a) All $\mathbb{U}$ units. (b) $\mathbb{U}$-$\mathbb{C}$ Transformation.

(a)

(b)

In FDP_{RI}, r probability cells from each action are initially composed as a Combination ($\mathbb{C}$) unit in Fig. 3.4 (a). n $\mathbb{C}$ units are initialized and equivalent to the one for DP_{RI} in Fig. 3.3 (a). When an update occurs, a $\mathbb{C}$ unit is transformed to a Unique ($\mathbb{U}$) unit which is composed of r probability cells of α_1 as shown in Fig. 3.4 (b). This updating result is equivalent to the one for DP_{RI} as shown in Fig. 3.3 (b).

Thus, the computational complexity to update the probability is reduced from $O(r)$ to $O(1)$ in FDP_{RI} because a type transformation from a $\mathbb{C}$ unit to a $\mathbb{U}$ unit can finish the probability updating.

In FDP_{RI}, n $\mathbb{C}$ units are initialized. When the probability update continues, the $\mathbb{C}$ units can be used up and transformed into the $\mathbb{U}$ units. Fig. 3.5 shows that r $\mathbb{U}$ units can be transformed into r $\mathbb{C}$ units, and they are equivalent. In this way, the transformed $\mathbb{C}$ units can be available for the subsequent fast probability update.

An action α_i is *active* if its state probability $p_i(t)$ is non-zero. When the state probability of an active action turns to be zero as shown in Fig. 3.6 (a), Re-organization is triggered as shown in Fig. 3.6 (b). Re-organization lets $\tilde{r} = \tilde{r} - 1$ and sets this action as non-active state where $\tilde{r}$ is number of the active actions and initially $\tilde{r} = r$. Hence, the step size turns to be $\Delta = \frac{1}{\tilde{r}n}$. The step size of the update scheme in FDP_{RI} for $p_m(t)$ "increases" because more and more state probabilities turn to be zero such that the number of the active actions decreases with the learning process. It means that the step size is "accelerated" along with the learning process. It is reasonable because the number of times for which active actions are selected increases, and the reward probability estimation of active actions becomes more and more accurate along with the learning process.

Then the $\mathbb{U}$-$\mathbb{C}$ transformation is called to transform $\mathbb{U}$ units into $\mathbb{C}$ units as possible for the subsequent fast probability updating. As shown in Fig. 3.6 (c), $\tilde{r} = 2$ $\mathbb{C}$

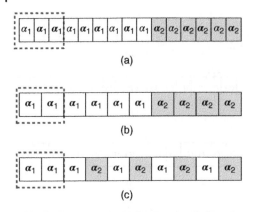

(a)

(b)

(c)

Figure 3.6 FDP$_{RI}$ re-organization using $r = 3$, $n = 5$ and $\tilde{r} = 2$. (a) α_3's probability turns to be zero. (b) Re-organization. (c) $\mathbb{U}$-$\mathbb{C}$ transformation.

units can be transformed by using one $\mathbb{U}$ unit of α_1 and one $\mathbb{U}$ unit of α_2. Therefore, $\mathbb{U}$-$\mathbb{C}$ transformation is performed twice at most and $2 \times \tilde{r} = 4$ $\mathbb{C}$ units are transformed.

In FDP$_{RI}$, an array F that stores the $\mathbb{U}$ and $\mathbb{C}$ units is initialized by n $\mathbb{C}$ units. With probability updating, some $\mathbb{C}$ units are consumed such that there are $\mathbb{U}$ and $\mathbb{C}$ units in F. For action selection, a random number $j = \lceil random \times n \rceil$ is obtained where random is sampled from the uniform distribution over $[0, 1]$. If the type of the j-th cell in F is $\mathbb{U}$, the owner of the corresponding $\mathbb{U}$ unit is chosen to interact with the environment. Else, one of the active actions is randomly chosen. In this way, the probability of choosing each action is proportional to the number of probability cells it has in array F and has a low computation complexity which is clearly independent of the number of actions.

Algorithm FDP$_{RI}$ presents the main procedure of the introduced algorithm. To avoid confusions, the notations used in FDP$_{RI}$ are listed in Table 3.4.

In FDP$_{RI}$, after the initialization, the chosen action is decided by the fast action selection in *Phase* 1. In *Phase* 2, the index of the estimated optimal action is found. In *Phase* 3, the state probability is updated if the response is positive. $\kappa \geq 1$ means that $\mathbb{C}$ units in F are available for fast probability update. It is notable that $\tilde{f}$ is used to indicate if there exist active actions whose probability turns to be zero. When $\kappa = 0$, no $\mathbb{C}$ unit is available for fast probability update when $\mathbb{U}$-$\mathbb{C}$ or Re-organization is called. $\mathbb{U}$-$\mathbb{C}$ produces a new $\mathbb{C}$ unit for the later fast probability update. If there exist active actions whose probability turns to be zero, $\tilde{r} = \tilde{r} - 1$ and F is reorganized for the later $\mathbb{U}$-$\mathbb{C}$ transformation and fast probability update. Finally, if max $\{P(t)\} \geq \hat{T}$ or $\tilde{r} = 1$, FDP$_{RI}$ is terminated.

As stated above, the fast action selection and the fast state probability update when $\mathbb{C}$ units are available have a low computational complexity which is independent of the number of actions. When $\mathbb{C}$ units are used up, ALGORITHM $\mathbb{U}$-$\mathbb{C}$ Transformation or ALGORITHM Re-organization is called to transform the $\mathbb{U}$ units into $\mathbb{C}$ units for the subsequent fast probability update.

FDP$_{RI}$: Fast Discretized Pursuit Reward-Inaction Algorithm
Input: Number of allowable actions r, resolution parameter n, action set A, environmental response set B, convergence threshold $\hat{T}$.
Output: Estimated optimal action α_m.
Initialize $\kappa = n; \varsigma = 0; \tilde{f} = 0; \tilde{r} = r; \tilde{c}(i) = 0, f(i = 0), active(i) = 1, \tilde{u}(i) = 0, \hat{\mathbb{I}}(i) = i, \check{\mathbb{I}}(i) = i, \forall i; F(j, 1) = 0, \tilde{e}(j) = j, j \in \{1, \ldots, n\}$;

Phase 1: Action Selection
$j = \lceil random \times n \rceil$;
If $F(j, 1) = 0$
  $i = \hat{\mathbb{I}}(\lceil random \times \tilde{r} \rceil)$;
Else
  $i = F(j, 2)$;
Endif
$\mathbb{H}_i(t) = \mathbb{H}_i(t) + \beta_i$
$\mathbb{G}_i(t) = \mathbb{G}_i(t) + 1$
$\tilde{d}_i = \mathbb{H}_i(t)/\mathbb{G}_i(t)$

Phase 2: Find the optimal estimated action
$m = max_i(d_i) \; \forall i, \; such \; that \; active(i) = 1$.

Phase 3: State Probability updating
If $\beta_m(t) = 1$
  If $\kappa \geq 1$
    inc(ς); inc($\tilde{u}(m)$); $\mathbb{I}(m, \tilde{u}(m)) = \tilde{c}(m)$;
    $\tilde{c}(m) = \tilde{e}(\kappa); F(\tilde{e}(\kappa), 1) = 1$;
    $F(\tilde{e}(\kappa), 2) = m$;
    dec(κ); $\hat{b} = de2bi(\tilde{f})$;
    If $size(\hat{b}) < \check{\mathbb{I}}(m)$
      $\tilde{f} = \tilde{f} + 2^{(\check{\mathbb{I}}(m)-1)}$;
    ElseIf $\hat{b}(index2(m)) = 0$
      $\tilde{f} = \tilde{f} + 2^{(\check{\mathbb{I}}(m)-1)}$;
    Endif
  Else If $\tilde{f} = (2^{\tilde{r}} - 1)$
      $\mathbb{U}\text{-}\mathbb{C}()$;
    Else
      Re-organization();
    Endif
  Endif
Endif
If $max\{P(t)\} \geq \hat{T}$ or $\tilde{r} = 1$, **Then** CONVERGE to the action α_m whose $p_m = max\{P(t)\}$. **Else** *Goto* **Phase 1**.
Endif
END

Table 3.4 Notations.

n	Resolution parameter, the larger, the higher resolution
r	Number of actions
$\tilde{r}$	Number of actions with non-zero probability and initially $\tilde{r} = r$
$\triangle = \frac{1}{\tilde{r}n}$	Step size of the probability update
$\tilde{A} = \{\alpha_1, ..., \alpha_{\tilde{r}}\}$	The set of actions whose probability is non-zero
$\mathbb{C}$	A unit of probability cells from all actions
$\mathbb{U}$	A unit of probability cells from a unique action
F	An array that stores the $\mathbb{U}$ and $\mathbb{C}$ units and initially by n $\mathbb{C}$ units
$F(j, 1)$	Indicating the type of $\mathbb{U}$ or $\mathbb{C}$ unit in F
$F(j, 2)$	Action index of the owner of the j-th $\mathbb{U}$ unit in F
κ	Number of $\mathbb{C}$ units in F and initialized as n
ς	Number of $\mathbb{U}$ units in F and initialized as 0
$\tilde{u}(i)$	Number of $\mathbb{U}$ units of the i-th action in F
$\tilde{c}(i)$	Current position index of the i-th action in F and initialized as . zero
$\mathbb{I}(i, j)$	Previous index before the j-th index of the i-th action in F
$\tilde{f}$	Indicates if there exist actions whose probabilities are equal to zero
$\hat{\mathbb{I}}$	Index in A of the i-th action in $\tilde{A}$ and initially $\hat{\mathbb{I}}(i) = i$
$\check{\mathbb{I}}$	Index in $\tilde{A}$ of the i-th action in A and initially $\check{\mathbb{I}}(i) = i$
$\tilde{e}(j)$	Position in F of the j-th available $\mathbb{C}$ unit

In $\mathbb{U}$-$\mathbb{C}$ Transformation, there exists a loop of $\tilde{r}$. However, after each time $\mathbb{U}$-$\mathbb{C}$ Transformation is performed, $\tilde{r}$ $\mathbb{C}$ cells are produced such that in the next $\tilde{r}$ iterations of LAs, $\mathbb{U}$-$\mathbb{C}$ Transformation are not called. Hence, the average computational complexity of $\mathbb{U}$-$\mathbb{C}$ Transformation is $O(1)$.

$\mathbb{U}$-$\mathbb{C}$ Transformation Algorithm

Input: Number of actions with non-zero probability $\tilde{r}$, number of $\mathbb{C}$ units κ, number of $\mathbb{U}$ units ς, array F, number of $\mathbb{U}$ units $\tilde{u}(i)$, current position index $\tilde{c}(i)$, position of available $\mathbb{C}$ unit $\tilde{e}(j)$, previous index $\mathbb{I}(i, j)$, index in A of the i-th action in $\tilde{A}$ $\hat{\mathbb{I}}$, index in $\tilde{A}$ of the i-th action in A $\check{\mathbb{I}}$.

Output: Number of $\mathbb{C}$ units κ, number of $\mathbb{U}$ units ς, array F, number of $\mathbb{U}$ units $\tilde{u}(i)$, current position index $\tilde{c}(i)$, position of available $\mathbb{C}$ unit $\tilde{e}(j)$, previous index $\mathbb{I}(i, j)$, index in A of the i-th action in $\tilde{A}$ $\hat{\mathbb{I}}$, index in $\tilde{A}$ of the i-th action in A $\check{\mathbb{I}}$.
$\tilde{f} = 0$;

---

for i=1:$\tilde{r}$

$\quad F(\tilde{e}(\kappa), 1) = 0; F(\tilde{e}(\kappa), 2) = 0;$

$\quad \text{inc}(\kappa); \tilde{e}(\kappa) = \tilde{c}(\hat{\mathbb{l}}(i));$

$\quad \tilde{c}(\hat{\mathbb{l}}(i)) = \mathbb{l}(\hat{\mathbb{l}}(i), \tilde{u}(\hat{\mathbb{l}}(i)));$

$\quad dec(\tilde{u}(\hat{\mathbb{l}}(i)));$

$\quad$**If** $(\tilde{u}(\hat{\mathbb{l}}(i))) > 0$

$\quad\quad \tilde{f} = \tilde{f} + 2^{i-1}$

$\quad$**Endif**

Endfor

$\varsigma = \varsigma - \tilde{r};$

END

For Algorithm Re-organization, after each time it is performed, at least one action turns to be non-active such that it is called at most $r - 1$ times. Hence, its average computation complexity for each iteration of LA is $O(1)$.

The computational complexities of the fast action selection, the fast state probability update, $\mathbb{U}$-$\mathbb{C}$ Transformation, and Re-organization are all $O(1)$. Hence, the FDP_{RI} performs both the action selection and the state probability update with a computational complexity which is independent of the number of actions.

Re-organization Algorithm

Input: Number of allowable actions r, resolution parameter n, number of actions with non-zero probability $\tilde{r}$, number of $\mathbb{C}$ units κ, number of $\mathbb{U}$ units ς, array F, number of $\mathbb{U}$ units $\tilde{u}(i)$, current position index $\tilde{c}(i)$, position of available $\mathbb{C}$ unit $\tilde{e}(j)$, previous index $\mathbb{l}(i,j)$, index in A of the i-th action in $\tilde{A}$ $\hat{\mathbb{l}}$, index in $\tilde{A}$ of the i-th action in A $\breve{\mathbb{l}}$.

Output: Number of actions with non-zero probability $\tilde{r}$, number of $\mathbb{C}$ units κ, number of $\mathbb{U}$ units ς, array F, number of $\mathbb{U}$ units $\tilde{u}(i)$, current position index $\tilde{c}(i)$, position of available $\mathbb{C}$ unit $\tilde{e}(j)$, previous index $\mathbb{l}(i,j)$, index in A of the i-th action in $\tilde{A}$ $\hat{\mathbb{l}}$, index in $\tilde{A}$ of the i-th action in A $\breve{\mathbb{l}}$.

$temp_{min} = n; \tilde{r} = 0;$

for i=1:r

$\quad$**If** $\tilde{u}(i) > 0$

$\quad\quad \text{inc}(\tilde{r}); \hat{\mathbb{l}}(\tilde{r}) = i; \breve{\mathbb{l}}(i) = \tilde{r};$

$\quad\quad$**If** $\tilde{u}(i) < temp_{min}$

$\quad\quad\quad temp_{min} = \tilde{u}(i);$

$\quad\quad$**Else**

$\quad\quad\quad active(i) = 0;$

$\quad\quad$**Endif**

$\quad$**Endif**

If $\check{r} = 1$
   break;
Endif
for j=1:$temp_{min}$
   $\mathbb{U}$-$\mathbb{C}$();
Endfor
END

3.2.3 Optimality Analysis

It is critically important to investigate if an algorithm is ε-optimal. Recently, a common flaw has been discovered by the authors of [20] in the previous proofs for the ε-optimality of almost all of the reported estimator algorithms. A new proof method that fixes this flaw is proposed in [20] by invoking the monotonicity property. Another proof method is then proposed in [21–23] to fix the flaw in the proof for the ε-optimality of the pursuit LA.

But instead of invoking the monotonicity property of the action probabilities, it examines their submartingale property and invokes the theory of Regular functions to prove the ε-optimality. This correct one is adopted in this book to prove the ε-optimality of our proposed algorithm.

The learning process of FDP$_{RI}$ can be divided into two phases: The first one is the sufficient sampling for all actions, in order to make the optimal action's rank accurate. During this phase, we cannot permit the state probability of any action to come to zero. This objective can be indeed achieved by improving the resolution n. The second phase is the pursuit of the estimated optimal action. Because the accuracy of the rank of the optimal action is high in the second phase, $p_m(t)$ keeps increasing such that the optimal action is never wiped out in probability $1 - \delta$. $1 - \delta$ is the probability that the optimal action correctly ranks in the first place during the whole learning process.

Theorem 3.1 *For each action α_i, $p_i(0) = \frac{1}{r}$. Then for any given constants $\delta > 0$ and $M < \infty$, there exist $n_0 < \infty$ and $t_0 < \infty$ such that under FDP$_{RI}$, for all learning parameters $n > n_0$ and all time $t > t_0$: for all $i \in \mathbb{Z}_r$,*

$$\Pr\{\mathbb{G}_i(t) \le M\} \le \delta, \text{ ensuring that } p_i(t') > 0 \text{ for } \forall\, t' \le t_0,$$

where $\mathbb{G}_i(t)$ is the number of times the ith action is chosen up to time t in any specific realization.

Proof: The proof of this theorem is analogous to the proof of the corresponding result for the DP_{RI} algorithm [12]. Consider FDP_{RI} and let $\tilde{r}(t) = |\{\alpha_i | p_i(t) \neq 0\}|$, where $|\cdot|$ denotes the cardinality of a set.

First, for an action, say α_i, we give a threshold of resolution parameter, denoted by $n(i)$, by which $p_i(t)$ remains non-zero if $t \leq t(i)$. Here, $t(i)$ is a threshold of time instant, which is to be defined later.

For action α_i, by virtue of the FDP_{RI}, at time $t \leq t(i)$:

$$
\begin{aligned}
p_i(t) &\geq \left(p_i(0) - \sum_{j=1\ldots t(i)} \frac{1}{\tilde{r}(t)n(i)} \right) \\
&\geq \left(p_i(0) - \sum_{j=1\ldots t(i)} \frac{1}{2n(i)} \right) \quad (3.16) \\
&= \left(p_i(0) - \frac{t(i)}{2n(i)} \right), \; i \in \mathbb{Z}_r. \quad (3.17)
\end{aligned}
$$

Then, under the premise that $(p_i(0) - t(i)/2n(i)) > 0$, we can arbitrarily set the value of $n(i)$ to be

$$
n(i) = r \cdot t(i). \quad (3.18)
$$

Second, we investigate the existence of $t(i)$, by which for any given constants $\delta > 0$ and $M < \infty$, when $t > t(i)$ and all learning parameters $n > n(i) = r \cdot t(i)$, it follow that $\Pr\{\mathbb{G}_i(t) \leq M\} \leq \delta$.

According to the inequalities (3.17) and (3.18), at time $t \leq t(i)$,

$$
\Pr\{\alpha_i \text{ is not chosen}\} \leq 1 - \frac{1}{2r} < 1. \quad (3.19)
$$

Based on the inequality (3.19), using a similar procedure to the proof of Theorem II in [12], we can get that for any given constants $\delta > 0$ and $M < \infty$, there exists such a threshold $t(i)$.

Since we can repeat this argument for all the actions, we can finally define t_0 and n_0 as follows:

$$
t_0 = \max_{i \in \mathbb{Z}_r} \{t(i)\}, \quad (3.20)
$$

$$
n_0 = \max_{i \in \mathbb{Z}_r} \{n(i)\} = \max_{i \in \mathbb{Z}_r} \{r \cdot t(i)\}, \quad (3.21)
$$

which can complete the proof. ∎

Based on Theorem 3.2.3, we can get the following theorem.

Theorem 3.2 *Given a* $\delta \in (0,1)$, *there exists a time instant* $t_0 < \infty$, *such that* $\Pr\{\bar{B}(t_0)\} = 1$. $\bar{B}(t_0)$ *is defined as follows:*

$$\bar{q}_j(t) = \Pr\left\{|\tilde{d}_j(t) - d_j| < \frac{w}{2}\right\},\tag{3.22}$$

$$\bar{q}(t) = \Pr\left\{|\tilde{d}_j(t) - d_j| < \frac{w}{2}, \forall j \in (1,2,...,r)\right\} = \prod_{j \in \mathbb{Z}_r} \bar{q}_j(t),\tag{3.23}$$

$$\tilde{B}(t) = \{\bar{q}(t) > 1 - \delta\}, \delta \in (0,1),\tag{3.24}$$

$$\bar{B}(t_0) = \left\{\bigcap_{t > t_0}\{\bar{q}(t) > 1 - \delta\right\},\tag{3.25}$$

where w is defined as the difference between the two highest reward probabilities.

Proof: This theorem proves the key condition $\bar{B}(t_0)$ for $p_m(t)_{t>t_0}$ with a $t_0 < \infty$ being a submartingale and fixes the flaw in the previous proofs reported in the literature where $\tilde{B}(t)$ is equivalent to $\bar{B}(t_0)$. This proof has been done in [21–23] and thus is omitted. ∎

Theorem 3.3 *Under FDP$_{RI}$, there exists a* $t_0 < \infty$ *such that the quantity* $p_m(t)_{t>t_0}$ *is a submartingale.*

Proof: From Theorem 3.1, at time instant t_0, $p_i(t_0) > 0$ for $\forall\, t' \leq t_0$. Define a time instant $\bar{t} \geq t_0$ such that:

$$\bar{t} = \sup\{t \mid p_m(t) \neq 0, t \geq t_0\}.$$

Note that here $\bar{t}$ is permitted to be ∞.

Next, we carry out the proof by analyzing two cases in terms of $\bar{t}$: (1) $\bar{t} = \infty$; (2) $\bar{t} < \infty$.

Case I: If $\bar{t} = \infty$, it means that $p_m(t) \neq 0$ for all $t \geq t_0$. According to the updating rule of FDP$_{RI}$ when $p_m(t) \neq 0$, it follows that:

$$E[p_m(t+1)|Q(t)] \geq \sum_{j=1...r} p_j(t)\left(d_j\left(\bar{q}(t)\left[p_m(t) + \frac{\tilde{r}(t) - 1}{n\tilde{r}(t)}\right]\right.\right.$$
$$\left.\left. + (1 - \bar{q}(t))\left[p_m(t) - \frac{1}{n\tilde{r}(t)}\right]\right) + (1 - d_j)p_m(t)\right)$$
$$= p_m(t) + \frac{\bar{q}(t)\tilde{r} - 1}{n\tilde{r}}\sum_{j=1...r} p_j(t)d_j.\tag{3.26}$$

Then,

$$E[p_m(t+1)|Q(t)] - p_m(t) \geq \frac{\bar{q}(t)\tilde{r} - 1}{n\tilde{r}}\sum_{j=1...r} p_j(t)d_j.\tag{3.27}$$

Invoking the termination condition for FDP$_{RI}$, in which we consider the learning process to have converged if max $\{Q(t)\} \geq \hat{T} = 1 - \varepsilon$ or $\tilde{r} = 1$. Therefore, during the learning process $\tilde{r}(t) \geq 2$. If we set the quantify $1 - \delta$ defined in Theorem 3.2 to be greater or equal than $\frac{1}{2}$, then for every single time instant subsequent to $t > t_0, \bar{q}(t) \geq 1 - \delta \geq \frac{1}{2}$,

$$E[p_m(t+1) - p_m(t)|Q(t)] \geq 0, \tag{3.28}$$

which implies that the quantity $p_m(t)_{t>t_0}$ is a submartingale.

Case II: If $\bar{t} < \infty$, the updating rule of FDP$_{RI}$ is determined in two disjoint phases.

Phase 1: When $t_0 \leq t < \bar{t}$, the updating rule of FDP$_{RI}$ for $p_m(t) \neq 0$ is adopted.

Phase 2: When $t \geq \bar{t}$, the updating rule of FDP$_{RI}$ for $p_m(t) = 0$ is adopted.

For Phase 1, we can use a similar analysis procedure to Case I. In Phase 2, by virtue of FDP$_{RI}$, it follows that $p_m(t) = 0$ for all $t \geq t_0$. Integrating both phases, for all $t > t_0$, the inequality (3.28) holds.

Combining two cases, we can get that the inequality (3.28) always holds, which completes the proof. ∎

Based on Theorem 3.3, by using the Martingale convergence theory [24], we can get the following corollary.

Corollary 3.1 *Under the FDP$_{RI}$, it holds that*

$$p_m(\infty) = 0 \text{ or } 1.$$

Theorem 3.4 *The FDP$_{RI}$ is ε-optimal in all stationary environments. More formally, let $\hat{T} = 1 - \varepsilon$ be a value arbitrarily close to 1, with ε being arbitrarily small. Then, given any δ, there exists a positive integer $n_0 < \infty$ and a time instant $t_0 < \infty$, such that for all learning parameters $n > n_0$ and for all $t > t_0$, $\bar{q}(t) > 1 - \delta > \frac{1}{2}$, $\Pr\{\lim_{t \to \infty} p_m(t) = 1\} = 1$.*

Proof: Denoting e_j as the unit vector with the jth element being 1, then our task is to prove the convergence probability

$$\Gamma_m(P) = \Pr\{p_m(\infty) = 1|P(0) = P\}$$
$$= \Pr\{p(\infty) = e_m|P(0) = P\} \to 1. \tag{3.29}$$

Defining a function $\Phi_m(P) = e^{-\tilde{x}_m p_m}$, where $\tilde{x}_m$ is a positive constant, and then define an operator $\bar{U}$ as:

$$\bar{U}(\Phi_m(P)) = E[\Phi_m(P(n+1))|P(n) = P]. \tag{3.30}$$

When $p_m(t) \neq 0$, it follows that:

$$\bar{U}(\Phi_m(P)) - \Phi_m(P) = E[\Phi_m(P(n+1))|P(n) = P] - \Phi_m(P)$$

$$= \sum_{j=1\ldots r} p_j(t) \cdot \left(d_j \cdot \left(\overline{q}(t) \cdot e^{-\tilde{x}_m \left(p_m(t) + \frac{\overline{r}(t)-1}{n\overline{r}(t)} \right)} \right. \right.$$

$$\left. \left. + (1 - \overline{q}(t)) \cdot e^{-\tilde{x}_m \left(p_m(t) - \frac{1}{n\overline{r}(t)} \right)} \right) + (1 - d_j) \cdot e^{-\tilde{x}_m p_m} \right)$$

$$- e^{-\tilde{x}_m p_m}$$

$$= \sum_{j=1\ldots r} p_j(t) \cdot \left(d_j \cdot \left(\overline{q}(t) \cdot e^{-\tilde{x}_m \left(p_m(t) + \frac{\overline{r}(t)-1}{n\overline{r}(t)} \right)} \right. \right.$$

$$\left. \left. + (1 - \overline{q}(t)) \cdot e^{-\tilde{x}_m \left(p_m(t) - \frac{1}{n\overline{r}(t)} \right)} \right) + (1 - d_j) \cdot e^{-\tilde{x}_m p_m} \right)$$

$$- \sum_{j=1\ldots r} p_j(t) \cdot e^{-\tilde{x}_m p_m}$$

$$= \sum_{j=1\ldots r} p_j(t) \cdot d_j \cdot e^{-\tilde{x}_m p_m(t)} \left(\overline{q}(t) \cdot e^{-\tilde{x}_m \left(\frac{\overline{r}(t)-1}{n\overline{r}(t)} \right)} \right.$$

$$\left. + (1 - \overline{q}(t)) \cdot e^{\tilde{x}_m \left(\frac{1}{n\overline{r}(t)} \right)} \right) - 1 \right)$$

$$= \sum_{j=1\ldots r} p_j(t) \cdot d_j \cdot e^{-\tilde{x}_m p_m(t)} \cdot e^{\frac{\tilde{x}_m}{n\overline{r}(t)}} \left(\overline{q}(t) \cdot \left(e^{\frac{-\tilde{x}_m}{n}} - 1 \right) \right.$$

$$\left. - e^{\left(\frac{-\tilde{x}_m}{n\overline{r}(t)} \right)} + 1 \right). \tag{3.31}$$

Let $\tilde{K} = \frac{\tilde{x}_m}{n}$ with $\tilde{K} \in (0, \infty)$. Then, let $\psi(\tilde{K}) = \frac{1 - e^{-\frac{1}{r} \cdot \tilde{K}}}{1 - e^{-\tilde{K}}}$. We can prove that the function $\psi(\tilde{K})$ is monotonically increasing as $\tilde{K}$ increases within $(0, \infty)$, which makes $\psi(\tilde{K})$ be subregular according to the following definitions of subregular/superregular functions in [24]:

A function $\chi \in \bar{C}(\hat{S}_r)$ is called *superregular* if

$$\chi(p) \geq \bar{U}(\chi(p)), \ \forall p \in \hat{S}_r, \tag{3.32}$$

where $\bar{C}(\hat{S}_r)$ denotes the class of all continuous functions mapping $\hat{S}_r \to \Re$ where $\Re$ is the real line.

Similarly, a function $\chi(\cdot)$ is called *regular* if

$$\chi(p) = \bar{U}(\chi(p)) \tag{3.33}$$

and *subregular*, if

$$\chi(p) \leq \bar{U}(\chi(p)) \tag{3.34}$$

over $p \in \hat{S}_r$.

Let $\psi(\tilde{K}_0) = \frac{1-e^{-\frac{1}{\tilde{r}}\cdot\tilde{K}_0}}{1-e^{-\tilde{K}_0}}$ and $\tilde{K}_0 = \frac{x_0}{n_0}$, we can get that if $n \geq n_0$ or $\tilde{x}_m \leq x_0$,

$$\bar{q}(t) \geq \psi(\tilde{K}_0) = \frac{1-e^{-\frac{1}{\tilde{r}}\cdot\tilde{K}_0}}{1-e^{-\tilde{K}_0}}, \tag{3.35}$$

such that

$$\bar{U}(\Phi_m(P)) - \Phi_m(P) \leq 0. \tag{3.36}$$

When $p_m(t) = 0$, it holds that:

$$\bar{U}(\Phi_m(P)) - \Phi_m(P) = 0. \tag{3.37}$$

Combining inequalities (3.36) and (3.37), we obtain that $\Phi_m(P)$ is superregular. According to the theory of regular functions [24], we have

$$\Gamma_m(P) \geq \phi(P) = \frac{1-e^{-\tilde{x}_m p_m(t)}}{1-e^{-\tilde{x}_m}}, \tag{3.38}$$

which meets the boundary conditions of (3.29). As the inequality (3.38) holds for every $\tilde{x}_m$ bounded by the inequality (3.35). In the inequality (3.35) if we take the greatest value of x_0, as $n_0 \to \infty$, $x_0 \to \infty$, such that $\Gamma_m(P) \to 1$ according to the definition of $\Gamma_m(P)$ in (3.29). We have thus proved that $\Pr\{\lim_{t\to\infty}p_m(t) = 1\} = 1$, showing that the FDP$_{RI}$ is ε-optimal. ∎

3.2.4 Simulation Studies

FDP$_{RI}$ is different from the classical DP$_{RI}$ in two aspects. One is that it performs both action selection and state probability update with a computational complexity independent of the number of actions. The other is the step size. In DP$_{RI}$, the step size is $\triangle = \frac{1}{rn}$, where r is the number of actions and n is a parameter. In FDP$_{RI}$, the step size is $\triangle = \frac{1}{\tilde{r}n}$, where $\tilde{r}$ is the number of active actions that is decreasing with its initial value being r. For the number of active actions decreases with the convergence, the step size of FDP$_{RI}$ increases while that of DP$_{RI}$ does not change. This section aims to compare the computation time and convergence speed of FDP$_{RI}$ and DP$_{RI}$ to show the superiority of the former. The other two famous pursuit scheme-based learning automata DGPA [13] and SE$_{RI}$ [14] are also included in these comparisons to help one well understand the advantages of the introduced FDP$_{RI}$.

The step size of FDP$_{RI}$ increases with the convergence, while that of DP$_{RI}$ remains unchanged. Consequently it speeds up the convergence significantly to be shown next. To compare their convergence speed, the popular benchmark test sets with ten actions are used to examine the performance of them. Table 3.5 shows the actions' reward probabilities for each environment. d_i is the reward probability of action i. The task of LA is to find the action with the highest

Table 3.5 Reward probability.

	d_1	d_2	d_3	d_4	d_5	d_6	d_7	d_8	d_9	d_{10}
E_1	**0.65**	0.50	0.45	0.40	0.35	0.30	0.25	0.20	0.15	0.10
E_2	**0.60**	0.50	0.45	0.40	0.35	0.30	0.25	0.20	0.15	0.10
E_3	**0.55**	0.50	0.45	0.40	0.35	0.30	0.25	0.20	0.15	0.10
E_4	**0.70**	0.50	0.30	0.20	0.40	0.50	0.40	0.30	0.50	0.20
E_5	0.10	0.45	**0.84**	0.76	0.20	0.40	0.60	0.60	0.50	0.30

Bold values are the max reward probability.

reward probability within fewest iterations. E_1–E_5 were used in [14]. E_4 and E_5 are the well-known benchmark environments and were used in [11–13, 15]. An algorithm is considered to have converged if the probability of choosing an action is not less than a threshold $\hat{T} = 0.999$ in all the performed tests. If the automaton converges to the action that has the highest reward probability, it is considered to have converged correctly.

As a common sense, the step size is a trade-off between the convergence accuracy and speed. Bigger n, i.e., higher resolution, leads to slower convergence and vice versa. γ is an additional parameter for the stochastic estimator in SE_{RI}. To compare the convergence speed of the contenders in a fair way, the minimum value of n is considered as the best value if it yields the fastest convergence given the accuracy equal to 99.5%. These best parameters are then chosen as the final parameter values used for the respective algorithms to compare their performances. Each result is obtained from 50 000 runs for each algorithm in each environment. Every algorithm samples all actions ten times each in order to initialize the estimated reward probability vector. These extra iterations are also included in the results presented in Table 3.6.

The increase of $p_m(t)$ in FDP_{RI} is slower in the early stages and faster in the late stages of the convergence than the ones in classical pursuit schemes. Table 3.6 shows that given an accuracy 99.5%, the FDP_{RI} converges much faster than the classical DP_{RI}, which indicates the increased step size of FDP_{RI} speeds up the convergence. It is reasonable because the number of times that active actions are selected increases such that the reward probability estimation of active actions becomes more and more accurate along with the learning process. Although DGPA and SE_{RI} converge faster, their computation complexity in phases 1 and 3 increases with the number of actions.

Figure 3.7 shows the CPU time for per iteration in FDP_{RI} and DP_{RI} where $n = 10$ and $n = 1000$ averaged from 2000 independent runs. The number of actionsis chosen from the set: {10, 50, 100, 500, 1000}. The reward probabilities of

Table 3.6 The average number of iterations required for convergence('ϱ' denotes the best parameter of contenders and 'ϑ' denotes the convergence speed).

	FDP_{RI}		DP_{RI}		DGPA		SE_{RI}	
	ϱ	ϑ	ϱ	ϑ	ϱ	ϑ	ϱ	ϑ
E_1	$n = 319$	864	$n = 298$	1088	$n = 27$	769	$n = 13, \gamma = 7$	334
E_2	$n = 722$	1933	$n = 681$	2569	$n = 59$	1366	$n = 32, \gamma = 11$	690
E_3	$n = 2943$	7963	$n = 3801$	12430	$n = 198$	4434	$n = 105, \gamma = 25$	1969
E_4	$n = 257$	658	$n = 205$	766	$n = 23$	589	$n = 7, \gamma = 6$	223
E_5	$n = 1015$	1879	$n = 1001$	2580	$n = 48$	1576	$n = 28, \gamma = 11$	731

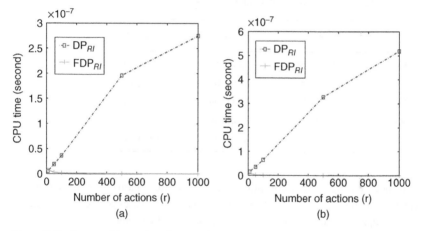

Figure 3.7 Computational time for each iteration in seconds. (a) $n = 10$. (b) $n = 1000$.

actions are set by taking equal intervals from $[0.1, 0.8]$ according to the actions numbers. All the experiments were conducted on a PC Intel (R) Pentium (R) CPU G630 at 2.70 GHZ with 4 Gb of RAM, running the Windows 7.

The curve of DP_{RI} shows the increase of CPU time for each iteration with the number of actions, while that of FDP_{RI} does not. Actually, the curve of FDP_{RI} decreases a little when the number of actions increases because the increase of the action count leads to more iterations for convergence such that the average CPU time of Algorithm Re-organization decreases.

Table 3.7 gives the average computational time for each updating consumed by the contenders. When the number of actions increases, the computational time increases monotonously and almost linearly for DP_{RI}, DGPA, and SE_{RI}. While the computational time for FDP_{RI} decreases a little with the increase of the number of

Table 3.7 The average computational time of contenders in seconds (*r* is the number of actions in the environment).

	FDP$_{RI}$	DP$_{RI}$	DGPA	SE$_{RI}$
r = 10	3.64E-09	5.69E-09	5.06E-09	4.23E-09
r = 50	1.26E-09	2.11E-08	2.50E-08	2.49E-08
r = 100	1.33E-09	4.16E-08	3.66E-08	4.94E-08
r = 500	1.68E-10	2.11E-07	3.23E-07	2.19E-07
r = 1000	1.91E-10	2.75E-07	5.25E-07	1.99E-07

Table 3.8 The average computational time of three phases of DP$_{RI}$ in seconds (*r* is the number of actions in the environment).

	Phase 1	Phase 2	Phase 3
r = 10	2.83E-09	6.04E-10	2.26E-09
r = 50	8.80E-09	1.00E-09	1.12E-08
r = 100	1.62E-08	1.41E-09	2.41E-08
r = 500	7.18E-07	3.74E-09	1.35E-07
r = 1000	1.05E-07	3.84E-09	1.67E-07

actions because the average computational time of Re-organization decreases due to the increasing iterations for more actions, which indicates the benefits brought by the FDP$_{RI}$ into the field of learning automata applications.

Table 3.8 presents the computational time for the three phases in traditional DP$_{RI}$. When the number of actions is large, the learning becomes slow because there are too many updates to be made at each iteration. The increased updates mostly result from the first and last, which indicates that phase 2 has the smallest computational complexity with the number of actions increasing such that the whole computational cost of DP_{RI} is reduced significantly because the computational cost of phases 1 and 3 is independent of the number of actions in the FDP$_{RI}$.

The updating step size can be made arbitrarily small to prevent the optimal action being wiped out. Indeed, these theoretical small steps are very conservative. In practice, the learning steps are usually far bigger than the theoretical steps [22]. Since FDP$_{RI}$ gets the accuracy equal to 99.5% when $n = 319$, let the step size decrease by increasing n from 369, 419, ..., to 1169 and run FDP$_{RI}$ on E_1. Table 3.9 shows that FDP$_{RI}$ can converge with an arbitrarily high probability as the learning step decreases at the expense of more computation.

Table 3.9 Increasing accuracy with decreasing step size.

ϱ	369	419	469	519	569	619	669	719
E_1	99.658%	99.782%	99.832%	99.918%	99.9288%	99.938%	99.966%	99.978%
ϱ	769	819	869	919	969	1019	1069	1169
E_1	99.994%	99.986%	99.988%	99.996%	99.990%	99.996%	99.998%	1

3.2.5 Summary

The discretized pursuit LA is the most popular LA as powerful tools of reinforcement learning in stochastic environments. However, when the number of actions is large, the learning becomes extremely slow because there are too many updates to be made at each iteration. This limitation prevents LAs from gaining large-scale action-oriented applications [25–28].

Compared with the classical DP_{RI}, the introduced fast discretized pursuit LA performs both action selection and state probability update with a computational complexity independent of the number of actions. Apart from its low computational complexity, it achieves faster convergence speed than the classical one when operating in stationary environments.

Compared with FLA [19], FDP_{RI} is ε-optimal and can be used in stationary environments. Moreover, it is based on the pursuit scheme such that it converges faster. Actually, if phase 2 of finding the estimated optimal action in FDP_{RI} is replaced by increasing the chosen action's probability without finding the estimated optimal action, we obtain a new fast scheme DL_{RI} which uses the LR scheme. If we do so, a fast LA based on DL_{RI} with the computational complexity $O(1)$ for all three phases can be obtained. However, the convergence speed is slower due to the linear scheme of DL_{RI} such that the overall computational amount increases due to the increasing number of iterations [9].

Compared with the hierarchical systems of learning automata [17, 18], the computational complexity of the probability updating is reduced from $O(\log r)$ in hierarchical automata to $O(1)$ in FDP_{RI}. Hence, the introduced fast and ε-optimal discretized pursuit learning automaton represents a significant advancement to the state-of-the-art in learning automata.

3.3 Exercises

1 What are the basic phases in a iteration of a discretized pursuit learning automaton?

2 What are the competitive characteristics of LELA?

3 Which phase does the LELA differentiate from other variants of learning automata?

4 What affects the convergence speed and accuracy?

5 What is the difference of update scheme among LELA, DGPA, and DP_{RI}?

6 How does LELA improve the convergence speed?

7 Design an LELA incorporating with stochastic estimator.

8 How to design a simulation to evaluate the performance of learning automata?

9 What is the difference between $LELA_S$ and $LELA_R$?

10 When does the $\mathbb{U}$-$\mathbb{C}$ transformation occur in FDP_{RI}?

11 Compute the computational complexity of FDP_{RI}.

12 What is the difference between FDP_{RI} and DP_{RI}?

13 Assume there are three actions and the resolution $n = 2$, the first action is always chosen and the environment always rewards it. Please give the update process of state probability using FDP_{RI}.

14 Try to devise a FLA based on LELA such that its state probability updates with a computational complexity independent of the number of actions.

References

1 Y. Wan, J. Qin, X. Yu, T. Yang, and Y. Kang, "Price-based residential demand response management in smart grids: A reinforcement learning-based approach," *IEEE/CAA Journal of Automatica Sinica*, vol. 9, no. 1, pp. 123–134, 2022.

2 Z. Cao, C. Lin, M. Zhou, and J. Zhang, "Surrogate-assisted symbiotic organisms search algorithm for parallel batch processor scheduling," *IEEE/ASME Transactions on Mechatronics*, vol. 25, no. 5, pp. 2155–2166, 2020.

3 Z. Cao, C. Lin, and M. Zhou, "A knowledge-based cuckoo search algorithm to schedule a flexible job shop with sequencing flexibility," *IEEE Transactions on Automation Science and Engineering*, vol. 18, no. 1, pp. 56–69, 2021.

4 J. Zhang, C. Wang, and M. Zhou, "Last-position elimination-based learning automata," *IEEE Transactions on Cybernetics*, vol. 44, no. 12, pp. 2484–2492, 2014.

5 V. I. Varshavskii and I. P. Vorontsova, "On the behavior of stochastic automata with variable structure," *Automatika i Telemekhanika (USSR)*, vol. 24, pp. 327–333, 1963.

6 M. L. Tsetlin, "On the behavior of finite automata in random media," *Automatika i Telemekhanika*, vol. 22, pp. 1345–1354, Oct 1961.

7 M. L. Tsetlin, "Automaton theory and the modeling of biological systems," New York: Academic, 1973.

8 M. A. L. Thathachar and B. J. Oommen, "Discretized reward-inaction learning automata," *Journal of Systemics, Cybernetics and Informatics Science*, vol. 2, pp. 24–29, 1979.

9 M. A. L. Thathachar and P. S. Sastry, "Estimator algorithms for learning automata," in *Proceedings of the Platinum Jubilee Conference on Systems and Signal Processing*, vol. Bangalore, India: Department of Electrical Engineering, December 1986, p. Indian Institute of Science.

10 M. A. L. Thathachar and P. S. Sastry, "A class of rapidly converging algorithms for learning automata," in *Proceedings of the IEEE International Conference on Cybernetics and Society*, Bombay, India, 1984, pp. 602–606.

11 B. J. Oommen and M. Agache, "Continuous and discretized pursuit learning schemes: Various algorithms and their comparison," *IEEE Transactions on Systems, Man, and Cybernetics, Part B (Cybernetics)*, vol. 31, no. 3, pp. 277–287, June 2001.

12 B. J. Oommen and J. K. Lanctot, "Discretized pursuit learning automata," *IEEE Transactions on Systems, Man, and Cybernetics*, vol. 20, pp. 931–938, July/Aug 1990.

13 M. Agache and B. J. Oommen, "Generalized pursuit learning schemes: New families of continuous and discretized learning automata," *IEEE Transactions on Systems, Man, and Cybernetics*, vol. 32, no. 6, pp. 738–749, 2002.

14 M. S. Georgios I. Papadimitriou, and A. S. Pomportsis, "A new class of ε-optimal learning automata," *IEEE Transactions on Systems, Man, and Cybernetics*, vol. 34, no. 1, pp. 246–254, 2004.

15 M. A. L. Thathachar and P. S. Sastry, "A new approach to the design of reinforcement schemes for learning automata," *IEEE Transactions on Systems, Man, and Cybernetics*, vol. SMC-15, pp. 168–175, Jan/Feb 1985.

16 J. Zhang, C. Wang, and M. Zhou, "Fast and epsilon-optimal discretized pursuit learning automata," *IEEE Transactions on Cybernetics*, vol. 45, no. 10, pp. 2089–2099, 2015.

17 M. A. L. Thathachar and K. R. Ramakrishnan, "A hierarchical system of learning automata," *IEEE Transactions on Systems, Man, and Cybernetics*, vol. 11, no. 3, pp. 236–241, 1981.

18 G. I. Papadimitriou, "Hierarchical discretized pursuit nonlinear learning automata with rapid convergence and high accuracy," *IEEE Transactions on Knowledge and Data Engineering*, vol. 6, no. 4, pp. 654–659, 1994.

19 M. S. Obaidat, G. I. Papadimitriou, and A. S. Pomportsis, "Efficient fast learning automata," *Information Sciences*, vol. 157, pp. 121–133, 2003.

20 M. Ryan and T. Omkar, "On ε-optimality of the pursuit learning algorithm," *Journal of Applied Probability*, vol. 49, no. 3, pp. 795–805, 2012.

21 X. Zhang, O.-C. Granmo, B. J. Oommen, and L. Jiao, "On using the theory of regular functions to prove the ε-optimality of the continuous pursuit learning automaton," in *Proceedings of International Conference on Industrial and Engineering Applications of Artificial Intelligence and Expert Systems*, June 2013, pp. 262–271.

22 X. Zhang, O.-C. Granmo, B. J. Oommen, and L. Jiao, "A formal proof of the ε-optimality of absorbing continuous pursuit algorithms using the theory of regular functions," *Applied Intelligence*, vol. 41, pp. 974–985, 2014.

23 X. Zhang, B. J. Oommen, O.-C. Granmo, and L. Jiao, "Using the theory of regular functions to formally prove the ε-optimality of discretized pursuit learning algorithms," in *Modern Advances in Applied Intelligence*. Springer, 2014, pp. 379–388.

24 K. S. Narendra and M. A. L. Thathachar, "Learning automata: An introduction," Englewood Cliffs, NJ: Prentice-Hall, 1989.

25 X. Guo, M. Zhou, A. Abusorrah, F. Alsokhiry, and K. Sedraoui, "Disassembly sequence planning: A survey," *IEEE/CAA Journal of Automatica Sinica*, vol. 8, no. 7, pp. 1308–1324, 2021.

26 L. Huang, M. Zhou, K. Hao, and H. Han, "Multirobot cooperative patrolling strategy for moving objects," *IEEE Transactions on Systems, Man, and Cybernetics: Systems*, vol. 53, no. 5, pp. 2995–3007, 2023.

27 Y. Tang, M. Zhou, and M. Gao "Fuzzy-Petri-Net based disassembly planning considering human factors," *IEEE Transactions on Systems, Man, and Cybernetics: Part A*, vol. 36, no. 4, pp. 718–726, 2006.

28 H. Yuan and M. Zhou, "Profit-Maximized collaborative computation offloading and resource allocation in distributed cloud and edge computing systems," *IEEE Transactions on Automation Science and Engineering*, vol. 18, no. 3, pp. 1277–1287, 2021.

4

Application-Oriented Learning Automata

Learning automata (LAs) have many variants for various application fields. Sometimes, these variants abstract characters from specific applied problems and own respective different learning schemes. In this chapter, two special variants are presented to show the flexible learning schemes in the LA field and inspire the readers to develop more vigorous and creative learning schemes from LA applications. One is to perform the discovery and tracking of spatiotemporal event patterns, and the other is to conduct the stochastic searching on a line.

4.1 Discovering and Tracking Spatiotemporal Event Patterns

Events are omnipresent in the real world [1–4]. For all the scope of events, they can be divided into two classes, i.e., Stochastically Episodic (SES) and Stochastically Non-Episodic (SNE) ones. For the latter, it is customary for one to model the behavior of accidents, telephone calls, network failures, etc., by using their respective probability distributions, which means that their occurrences obey a stochastic pattern. For the former, we can neither predict its occurrence nor model it as a pattern, such as earthquakes and nuclear explosions. A stream of SNE events can be represented as a spatiotemporal pattern. This pattern describes these events' occurrences over time at a specific location [5]. Discovering and tracking spatiotemporal event patterns have many applications. For example, in a smart-home project, a set of spatiotemporal pattern LAs can be used to monitor a user's repetitive activities, by which the home's automaticity can be promoted while some of their burdens can be reduced.

As a matter of fact, a stream of events has its own physical significance and appears in our daily lives. For example, a daily group meeting is a common phenomenon in a company held every day at a meeting room, such that a 0/1 (nonoccurrence/occurrence) sequence can express whether the meeting is held

Learning Automata and Their Applications to Intelligent Systems, First Edition.
JunQi Zhang and MengChu Zhou.

at the meeting room every day or not. Similar instances include: (i) The Internet of Things: a 0/1 sequence expresses whether a server receives a sensor's signal; (ii) The ambient assisted living: a 0/1 sequence expresses whether an elderly resident goes to the kitchen or not. Hence, discovering and tracking spatiotemporal event patterns is to determine if a spatiotemporal pattern exists in a given event stream, namely, perceiving a pattern change efficiently in a specific time and location under a noisy environment. Note that a pattern's change means that a pattern switches from existence to nonexistence or vice versa.

Example 4.1 The following example is due to [5, 6]. In a smart home, a user's daily activities usually generate multiple patterns in a finite activity space. A spatiotemporal pattern recognition framework considers a discrete finite world consisting of $\dot{m}$ spatial location primitives $\dot{L} = \{\dot{l}_1, \dot{l}_2, \dots, \dot{l}_{\dot{m}}\}$ and $\dot{n}$ discrete-time primitives $\dot{T} = \{\dot{t}_1, \dot{t}_2, \dots, \dot{t}_{\dot{n}}\}$ with a proper granularity to describe the multiple patterns. Let $\dot{L} = \{$"University," "Office," "Gym"$\}$, and $\dot{T} = \{$"Mondays," "Tuesdays," "Weekends," "Daily"$\}$. The spatiotemporal pattern space equals a 12-element set $\dot{P} = \{$"Mondays at University," "Mondays at Office," "Mondays at Gym," "Tuesdays at University," "Tuesdays at Office," "Tuesdays at Gym," "Weekends at University," "Weekends at Office," "Weekends at Gym," "Daily at University," "Daily at Office," "Daily at Gym"$\}$. Each element in this set is seen as a hypothesis such that this set corresponds to a hypothesis set $\dot{H} = \{\dot{h}_1, \dot{h}_2, \dots, \dot{h}_{\dot{r}}\}$, where $\dot{r} = \dot{m} \times \dot{n}$.

Some algorithms for spatiotemporal event pattern recognition in a noisy dynamic environment are based on Fixed Structure Stochastic Automata whose state transition function is fixed and predesigned to guarantee noise immunity. However, such a design is conservative because it needs continuous and identical feedback to converge, thus resulting in its slow convergence in practice. In many real-life applications, such as ambient assisted living, continuous nonoccurrences of an elderly resident's routine activities should be treated with an Alert as quickly as possible. Also, no Alert should be output even for some occurrences to diminish the effects caused by noise.

Clearly, confronting a pattern's change, slow speed and low accuracy may cause a user's life security issue. This section first introduces a fast and accurate learning automaton based on Variable Structure Stochastic Automata named Adaptive Tunable Spatio-Temporal Pattern Learning Automata (AT-STPLA) [7] to satisfy the realistic requirements for both speed and accuracy. Furthermore, bias toward Alert is necessary for elderly residents, while the existing STPLA can only support the bias toward "no Alert." Hence, an AT-STPLA-based method toward either Alert or "no Alert" is introduced to meet a user's specific bias requirements. The convergence proof of AT-STPLA and experiments on its performance are also given.

4.1.1   Background and Motivation

In the framework [5], the whole recognition system consists of numerous LAs. As a learning unit, every LA's performance determines the efficiency of the whole recognition system. Promoting these LAs' convergence rate and accuracy is crucial to the entire system's performance. Meanwhile, existing Fixed Structure Stochastic Automata (FSSA) contradicts that once the pattern changes, it requires continuous and identical rewards or penalties to jump into the opposite states, which decreases the convergence speed. This defect is due to their fixed structure. Their state transition strategy is rigid and unable to adapt to their environment dynamically.

In [5], Yazidi *et al.* present a Spatio-Temporal Pattern Learning Automaton (STPLA) to recognize a spatiotemporal pattern and arrange a single STPLA to conjecture a single hypothesis. Its learning result indicates whether the corresponding hypothesis is tenable or not (named Suppress or Notify). Imagine its application in a smart home project – a single resident has many probable routine activities. To learn every activity's occurrence or nonoccurrence, we should maintain a hypothesis set containing many hypotheses covering every possible pair of period and location. Undoubtedly, the scale of this set is enormous such that LAs with a small footprint are the best implementation of the framework in [5].

In [8], based on the framework of [5], Jiang *et al.* present Spatio-Temporal Pattern Tunable Fixed Structure Learning Automata (STP-TFSLA) to satisfy the bias preference toward Suppress in the scenario of Internet of Things. Such LA is a combination of STPLA and TFSLA [9]. Instead of changing the LAs' fixed structure discretely, it provides an intuitive parameter to the user, relaxing the relationship between balanced memory and bias. STP-TFSLA unilaterally enhances the accuracy and convergence rate when a pattern exists. On the other hand, its Alert capability drops when a pattern disappears.

The contents of this section are:

1) To discover and track spatiotemporal patterns through a noisy event stream, this section introduces a Variable Structure Stochastic Automaton (VSSA) instead of a fixed one to obtain faster and more accurate convergence by an adaptive state-transition step. The adaptive step is self-regulated along an event stream. Moreover, such a step makes the structure of this LA variable.
2) By considering a user's preference under different application scenarios, this chapter first provides an interface to decide if an LA's bias should be toward Notify (Alert), Suppress (no Alert), or Balance. We name this LA as an Adaptive Tunable Spatio-Temporal Pattern Learning Automaton (AT-STPLA).
3) Three sets of experiments are performed to demonstrate that the new LA's better performance to discover and track spatiotemporal patterns in both static and dynamic noisy environments than the state-of-the-art approaches.

Formally, an LA is formulated as a 5-tuple (A, B, Q, T, G), where $A = \{\alpha_1, \alpha_2, \ldots, \alpha_r\}$ is an input action set, $B = \{\beta_1, \beta_2, \ldots, \beta_u\}$ is an output action set, $Q = \{q_1(t), q_2(t), \ldots, q_v(t)\}$ is the internal state at instant t, T is the state transition function, i.e., $q(t + 1) = T(q(t), \alpha(t), \beta(t))$, and G is the output function for outputting $\alpha(t)$ according to $q(t)$.

In the former, T and G are immutable, which means these two maps do not change with time. FSSA mainly includes Tsetlin LA [10], Krinsky LA [11] (the prototype of STPLA in [5]), and TFSLA [9] (the prototype of STP-TFSLA in [8]) introduced in Chapter 2 of this book. In the latter, either T or G, or both, vary with time. This is why this kind of LAs is named as "Variable" Structure Stochastic Automata. A probability vector usually represents their state.

4.1.2 Spatiotemporal Pattern Learning Automata

In [6], Rashidi and Cook introduced an interacting system to realize a smart home. In this system, the off-line Frequent and Periodic Activity Miner (FPAM) algorithm is introduced to find repetitive patterns in a resident's routine activities, including two phases of finding and evaluating candidate patterns.

Yazidi *et al.* [5] first formulated an STPLA-based spatiotemporal pattern recognition framework, in which an LA was seen as a basal learning unit, as shown in Fig. 4.1. FPAM is modified in an on-line manner. Besides its poor performance, it is no longer applicable in this framework because if every single hypothesis possesses one sliding window mechanism, redundant storage burden becomes inevitable. STPLA is a 5-tuple where

1) $A = \{Notify, Suppress\}$.
2) $B = \{Reward, Penalty\}$.
3) $Q = \{1, 2, \ldots, N_1, N_1 + 1, \ldots, N_1 + N_2 + 1\}$.
4) $T(0)$ is the transition matrix when the feedback is *Reward* and $T(1)$ is the transition matrix when the feedback is *Penalty*. They are represented as

$$T(0) = \begin{pmatrix} 0 & 1 & 0 & \cdots & 0 & 0 & 0 & \cdots & 0 \\ 0 & 0 & 1 & \cdots & 0 & 0 & 0 & \cdots & 0 \\ \vdots & \vdots & \ddots & \ddots & \vdots & \vdots & \vdots & & \vdots \\ 0 & 0 & \cdots & 0 & 1 & 0 & 0 & \cdots & 0 \\ 0 & 0 & \cdots & 0 & 1 & 0 & 0 & \cdots & 0 \\ 0 & 0 & \cdots & 0 & 1 & 0 & 0 & \cdots & 0 \\ \vdots & \vdots & \vdots & \vdots & \vdots & \vdots & \ddots & \ddots & \vdots \\ 0 & 0 & \cdots & 0 & 1 & 0 & \cdots & 0 & 0 \\ 0 & 0 & \cdots & 0 & 1 & 0 & 0 & \cdots & 0 \end{pmatrix},$$

$$T(1) = \begin{pmatrix} 1 & 0 & 0 & \cdots & 0 & 0 & 0 & \cdots & 0 \\ 1 & 0 & 0 & \cdots & 0 & 0 & 0 & \cdots & 0 \\ \vdots & \vdots & \ddots & \ddots & \vdots & \vdots & \vdots & & \vdots \\ 1 & 0 & \cdots & 0 & 0 & 0 & 0 & \cdots & 0 \\ 0 & 0 & \cdots & 0 & 0 & 1 & 0 & \cdots & 0 \\ 0 & 0 & \cdots & 0 & 0 & 0 & 1 & \cdots & 0 \\ \vdots & \vdots & \vdots & \vdots & \vdots & \vdots & \ddots & \ddots & \vdots \\ 0 & 0 & \cdots & 0 & 0 & 0 & \cdots & 0 & 1 \\ 1 & 0 & \cdots & 0 & 0 & 0 & 0 & \cdots & 0 \end{pmatrix}.$$

5) $G(q(t)) = \begin{cases} \textit{Notify}, & \text{if } q(t) \in \mathbb{S}_1 \\ \textit{Suppress}, & \text{if } q(t) \in \mathbb{S}_2 \end{cases}$,

where $q(t)$ is the state at instant t and $\mathbb{S}_1 = \{1, 2, \ldots, N_1\}$, $\mathbb{S}_2 = \{N_1 + 1, N_2 + 2, \ldots, N_1 + N_2 + 1\}$.

Such LA does not need any prior knowledge to adapt to a pattern's change and can decrease the storage requirement. Additionally, STPLA can work in a noisy environment, which includes two types of errors, as defined in [5] and recalled as follows:

1) *Omission Error*: An event is supposed to have occurred according to the pattern but actually not. It has the nature of an Stochastic EnVironment (SEV) event. $\tilde{q}$ represents the probability of omission error occurrence. For example, in a company's busy season, daily meetings are held in a certain meeting room, which can be seen as a pattern. However, in several days, meetings can be cancelled for unpredictable reasons, such that these nonoccurrences of meetings are defined as omission errors in a daily pattern.

2) *Inclusion Error*: An event that occurs is not a part of a periodic pattern but rather arises sporadically and spontaneously. It also obeys the nature of an SES event. $\tilde{p}$ represents the probability of inclusion error occurrence. For example, in an idle season of a company, daily meetings should be cancelled, which means a pattern does not exist. Nevertheless, in several days, some random tasks lead the meetings to be held in a certain meeting room, such that these occurrences of meetings are defined as inclusion errors when a pattern does not exist.

Example 4.2   From Fig. 4.1, in a Pattern Evaluation Phase, if a pattern exists and an LA at state 1 intends to converge correctly, it actually needs continuous N_1 rewards. Let the omission error probability q equal 0.2 and $N_1 = 12$, the convergence probability equal $(1 - q)^{12} = 0.0687$ after 12 instants, which is too small to lead to a convergence.

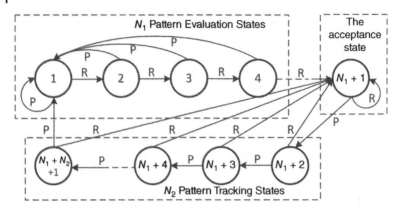

Figure 4.1 State transition map and the output function of an STPLA Source: Adapted from [5].

Based on the analysis mentioned above, we conclude that the original state transition principles are conservative and should be more dynamic and adaptive. VSSA is a good choice to meet such requirements due to its flexible and variable state-transition rules. Furthermore, in the ambient assisted living environment, bias toward Notify is also necessary for the elders because any nonoccurrence of their routine activities should also be brought to medical attention.

STP-TFSLA [8] shown in Fig. 4.2 utilizes TFSLA's idea to make an STPLA inclined to Suppress. It relaxes the relationship between balanced memory and bias to overcome some real-world application problems such as data collection in IoT. Although STPLA and STP-TFSLA can discover and track patterns in noisy environments, their convergence needs to speed up to meet the many real-life

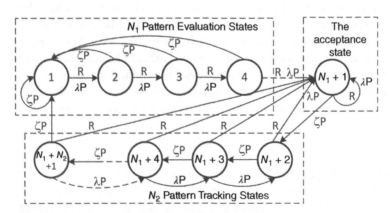

Figure 4.2 State transition map and the output function of an STP-TFSLA. Source: [8]/with permission of Elsevier.

application scenarios, such as ambient assisted living [12]. Their slow response speed confronting a pattern's change may endanger a human user's life.

4.1.3 Adaptive Tunable Spatiotemporal Pattern Learning Automata

This section introduces a VSSA named Adaptive Tunable Spatio-Temporal Learning Automaton (AT-STPLA), in which "adaptive" means its step size is self-regulated and "tunable" has the same implication as [8], meaning that its bias preference is adjustable.

An AT-STPLA can be described as a 5-tuple (A, B, Q, T, G), where A and B have the same meaning as those in STPLA. The state set $Q = \{1, 2, \ldots, N, N + 1, \ldots, 2N\}$. For these $2N$ states, we divide them into two symmetrical parts. The first part is called the Pattern Evaluation States containing states $\mathbb{S}_1 = \{1, \ldots, N\}$ and another part is called the Pattern Tracking States containing states $\mathbb{S}_2 = \{N + 1, \ldots, 2N\}$. Thus, we can define the map G as follows,

$$G(q(t)) = \begin{cases} Notify, & \text{if } q(t) \in \mathbb{S}_1 \\ Suppress, & \text{if } q(t) \in \mathbb{S}_2 \end{cases}$$

where $q(t)$ is a real-time state at instant t.

For AT-STPLA, feedbacks from an environment do not influence $q(t)$ directly, but first exerts influence on the state transition principle T. T, at instant t, can be represented by step size $\bar{s}(t)$, varying with time. The update principles of $\bar{s}(t)$ and $q(t)$ are described as follows.

If *Reward* is received,

$$\bar{s}(t + 1) = \begin{cases} \bar{s}(t) + 1, & \text{if } \bar{s}(t) < S \\ \bar{s}(t), & \text{if } \bar{s}(t) = S \end{cases}. \tag{4.1}$$

If *Penalty* is received,

$$\bar{s}(t + 1) = \begin{cases} \bar{s}(t) - 1, & \text{if } \bar{s}(t) > -S \\ \bar{s}(t), & \text{if } \bar{s}(t) = -S \end{cases}. \tag{4.2}$$

After $\bar{s}(t)$ is updated, so is Q via:

$$q(t + 1) = \begin{cases} q(t) + \bar{s}(t + 1), & \text{if } 1 \leq q(t) + \bar{s}(t + 1) \leq 2N \\ 2N, & \text{if } q(t) + \bar{s}(t + 1) > 2N \\ 1, & \text{if } q(t) + \bar{s}(t + 1) < 1, \end{cases} \tag{4.3}$$

In (4.1) and (4.2), S ($S > 0$) is a function of N,

$$S = \left\lceil \frac{\sqrt{16N - 7} - 1}{2} \right\rceil. \tag{4.4}$$

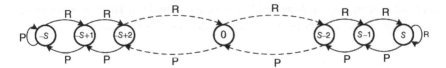

Figure 4.3 The state transition principle of $\bar{s}(t)$. R means *Reward* and P means *Penalty*. Due to the hypothesis of an AT-STPLA, an occurrence of an event corresponding to a *Reward* has probability $\bar{p}$. The nonoccurrence of an event corresponding to a *Penalty* has probability $\bar{q}$.

$\left\lceil (\sqrt{16N - 7} - 1)/2 \right\rceil$ is the maximum value of $\bar{s}(t)$ when $q(t)$ arrives at state $2N$ from a non-$2N$ state. $-\left\lceil (\sqrt{16N - 7} - 1)/2 \right\rceil$ is the minimum value of $\bar{s}(t)$ when $q(t)$ arrives at state 1 from a non-1 state. The state transition of $\bar{s}(t)$ is shown in Fig. 4.3 and (4.4) is deduced next.

As the matter of fact, step size $\bar{s}(t)$ plays an estimator-like role to a certain extent, which has the function of historical recording information to make an accurate decision. In spatiotemporal pattern recognition, LAs work in a dynamic environment. Historical information accumulation may become a hindrance to pattern tracking. Thus we set the step size $\bar{s}(t)$ in a reasonable range as in (4.1) and (4.2). $S = \left\lceil \left(\sqrt{16N - 7} - 1 \right)/2 \right\rceil$ is a suitable value to make this tradeoff.

An AT-STPLA's is initialized as follows.

1) The initial state $q(0)$ is set randomly among 1 to $2N$.
2) The initial step size $\bar{s}(0)$ is set as 0.

Example 4.3 When $N = 5$, an unfolded state transition principle with different T is listed in Fig. 4.4. But note that a countable and finite T does not mean that the structure is fixed because the transition principle T at time t is determined by $\bar{s}(t)$, while $\bar{s}(t)$ is time-varying. Here we give a trace of performance when a pattern exists and $q = 0.2$, in which $N = 5$, $S = \left\lceil (\sqrt{16 * 5 - 7} - 1)/2 \right\rceil = 4$, $q(0) = 1$, and $\bar{s}(0) = 0$.

Step 1: Current Step Size: 0, Current State: 1
Feedback: *Reward*
Next Step Size: $0 + 1 = 1$, Next State: $1 + 1 = 2$
Step 2: Current Step Size: 1, Current State: 2
Feedback: *Reward*
Next Step Size: $1 + 1 = 2$, Next State: $2 + 2 = 4$
Step 3: Current Step Size: 2, Current State: 4
Feedback: *Reward*
Next Step Size: $2 + 1 = 3$, Next State: $4 + 3 = 7$

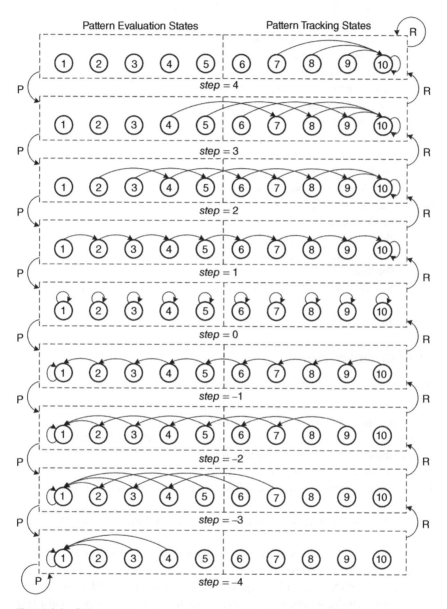

Figure 4.4 State transition map and the output function of an AT-STPLA. Note that when *step* = ±2, ±3, and ±4 there is no arrow on some state circles because those states cannot be reached. For example, when *step* = 3, current state must be 7, 8, 9, or 10 but not others.

Step 4: Current Step Size: 3, Current State: 7
 Feedback: *Penalty*
 Next Step Size: $3 - 1 = 2$, Next State: $7 + 2 = 9$
Step 5: Current Step Size: 1, Current State: 2
 Feedback: *Reward*
 Next Step Size: $2 + 1 = 3$, Next State: 10
Step 6: Current Step Size: 3, Current State: 10
 Feedback: *Reward*
 Next Step Size: $3 + 1 = 4$, Next State: 10
Step 7: Current Step Size: 4, Current State: 10
 Feedback: *Reward*
 Next Step Size: 4, Next State: 10
Step 8: Current Step Size: 4, Current State: 10
 Feedback: *Penalty*
 Next Step Size: $4 - 1 = 3$, Next State: 10
Step 9: Current Step Size: 3, Current State: 10
 Feedback: *Reward*
 Next Step Size: $3 + 1 = 4$, Next State: 10

Thus, this AT-STPLA converges successfully.

4.1.4 Optimality Analysis

Before all the analyses, we give the necessary definitions as follows.

1) If events follow a regular pattern, then $\tilde{p}$ is occurrence probability and $\tilde{q}$ is assimilated to omission noise probability.
2) If events do not follow a regular pattern, then $\tilde{p}$ is assimilated to inclusion noise probability and $\tilde{q}$ is nonoccurrence probability.
3) For convenience, let $\tilde{a} = \tilde{p}/\tilde{q}$.
4) $\bar{s}(t)$ means the current step size at instant t and $q(t)$ means the current state at instant t.
5) $\hat{\Gamma}$ is the ceiling of step size, which can be calculated by (4.4), such that $\bar{s}(t) \in \{-S, \ldots, -1, 0, 1, \ldots, S\}$.
6) $\mathbb{S}_1 = \{1, \ldots, N\}$ is the set of Pattern Evaluation States. $\mathbb{S}_2 = \{N + 1, \ldots, 2N\}$ is the set of Pattern Tracking States.
7) $\mathbb{P}_1$ is the Alert (Notify) probability, which means a probability of $q(t) \in \mathbb{S}_1$ after t instant; $\mathbb{P}_2$ is the Suppress probability, which means a probability that of $q(t) \in \mathbb{S}_2$ after instant t. Obviously, $\mathbb{P}_1 + \mathbb{P}_2 = 1$.

Theorem 4.1 *When $q(t)$ arrives at state 2N from a non-2N state, the maximum value of $\bar{s}(t)$ is $\left\lceil (\sqrt{16N - 7} - 1)/2 \right\rceil$. When $q(t)$ arrives at state 1 from a non-1 state, the minimum value of $\bar{s}(t)$ is $- \left\lceil (\sqrt{16N - 7} - 1)/2 \right\rceil$.*

Proof: When $q(t)$ arrives at state $2N$ from a non-$2N$ state, under one condition $\bar{s}(t)$ can reach the maximum value $\hat{x}$ – starting at state 1 and receiving continuous rewards. Its step size varies from 0 to S step by step and satisfies

$$(1 + 2 + \cdots + \hat{x}) + 1 = \frac{(\hat{x} + 1)\hat{x}}{2} + 1 \geq 2N. \tag{4.5}$$

Solving (4.5), we obtain $\hat{x} \leq \left\lfloor (-\sqrt{16N - 7} - 1)/2 \right\rfloor$ or $\hat{x} \geq \left\lceil (\sqrt{16N - 7} - 1)/2 \right\rceil$. Adopting the minimum positive integer, the maximum value of $\bar{s}(t)$ is represented as $S = \left\lceil (\sqrt{16N - 7} - 1)/2 \right\rceil$.

In the same way, when $q(t)$ arrives at state 1 from a non-1 state, if $\bar{s}(t)$ can reach the minimum value $-\hat{x}$ ($\hat{x} > 0$), $\hat{x}$ satisfies

$$2N + ((-1) + (-2) + \cdots + (-\hat{x})) = 2N - \frac{(\hat{x} + 1)\hat{x}}{2} \leq 1. \tag{4.6}$$

Solving (4.6), we obtain $-\hat{x} \geq \left\lceil \left(\sqrt{16N - 7} + 1 \right)/2 \right\rceil$ or $-\hat{x} \leq -\left\lceil \left(\sqrt{16N - 7} - 1 \right)/2 \right\rceil$. The minimum value $-\hat{x}$ should adopt the maximum negative integer and it is represented as $-S = -\left\lceil \left(\sqrt{16N - 7} - 1 \right)/2 \right\rceil$.

Hence the Theorem 4.1. ∎

Theorem 4.2 *When $t \to \infty$, let $\bar{\pi}_{\bar{s}(t)}$ be the probability of $\bar{s}(t)$. Then*

$$\bar{\pi}_{\bar{s}(t)} = \frac{1 - \tilde{a}}{1 - \tilde{a}^{2S+1}} \tilde{a}^{\bar{s}(t)}. \tag{4.7}$$

Proof: As shown in Fig. 4.3, $\bar{s}(t)$ follows a Markov chain model. Transition probability matrix $\overline{M}$ is:

$$\overline{M} = \begin{pmatrix} \tilde{q} & \tilde{p} & 0 & \cdot & 0 & 0 & \cdot & \cdot & 0 & 0 \\ \tilde{q} & 0 & \tilde{p} & \cdot & 0 & 0 & \cdot & \cdot & 0 & 0 \\ 0 & \tilde{q} & 0 & \cdot & 0 & 0 & \cdot & \cdot & 0 & 0 \\ \cdot & \cdot & \cdot & \cdot & \cdot & \cdot & \cdot & \cdot & \cdot & \cdot \\ 0 & 0 & 0 & \cdot & 0 & \tilde{p} & \cdot & \cdot & 0 & 0 \\ 0 & 0 & 0 & \cdot & \tilde{q} & 0 & \tilde{p} & \cdot & 0 & 0 \\ 0 & 0 & 0 & \cdot & \cdot & q & 0 & \cdot & 0 & 0 \\ \cdot & \cdot & \cdot & \cdot & \cdot & \cdot & \cdot & \cdot & \cdot & \cdot \\ 0 & 0 & 0 & \cdot & 0 & 0 & \cdot & \cdot & 0 & \tilde{p} \\ 0 & 0 & 0 & \cdot & 0 & 0 & \cdot & \cdot & \tilde{q} & \tilde{p} \end{pmatrix}. \tag{4.8}$$

We aim to compute the stationary (or equilibrium) probability of the Markov chain being at state $\bar{s}(t)$. The chain is ergodic and the eigenvector of $\overline{M}^T$ corresponding gives the limiting probability vector to the eigenvalue unity. Let the vector be $\overline{\Pi} = (\pi_{-S}, \ldots, \pi_{-1}, \pi_0, \pi_1, \ldots, \pi_S)^T$. Then $\overline{\Pi}$ satisfies

$$\overline{M}^T \overline{\Pi} = \overline{\Pi}, \tag{4.9}$$

which concludes the following equations:

$$
\begin{aligned}
\overline{\pi}_{-S} &= \tilde{q}\overline{\pi}_{-S} + \tilde{q}\overline{\pi}_{-S+1} \\
\overline{\pi}_{-S+1} &= \tilde{p}\overline{\pi}_{-S} + \tilde{q}\overline{\pi}_{-S+2} \\
\overline{\pi}_{-S+2} &= \tilde{p}\overline{\pi}_{-S+1} + \tilde{q}\overline{\pi}_{-S+3} \\
&\vdots \\
\overline{\pi}_{S-2} &= \tilde{p}\overline{\pi}_{S-3} + \tilde{q}\overline{\pi}_{S-1} \\
\overline{\pi}_{S-1} &= \tilde{p}\overline{\pi}_{S-2} + \tilde{q}\overline{\pi}_{S} \\
\overline{\pi}_{S} &= \tilde{p}\overline{\pi}_{S-1} + \tilde{p}\overline{\pi}_{S}
\end{aligned}
\tag{4.10}
$$

From (4.10), we obtain for $-S \le k < S$

$$\overline{\pi}_{k+1} = \tilde{a}\overline{\pi}_k. \tag{4.11}$$

Thus, $\overline{\pi}_k$ forms a geometric progression, of which the common ratio is $\tilde{a}$. $\bar{s}(t)$ is the $\left(S + 1 + \bar{s}(t)\right)$-th item, which concludes

$$\overline{\pi}_{\bar{s}(t)} = \overline{\pi}_{-S}\tilde{a}^{S+\bar{s}(t)}. \tag{4.12}$$

Meanwhile, the sum of all the elements in vector $\overline{\Pi}$ equals 1, i.e.,

$$
\begin{aligned}
\sum_{-S}^{S} \overline{\pi}_k &= \frac{1-\tilde{a}^{2S+1}}{1-\tilde{a}} \overline{\pi}_{-S} \\
&= 1.
\end{aligned}
\tag{4.13}
$$

Therefore, from (4.13) we obtain

$$\overline{\pi}_{-S} = \frac{1-\tilde{a}}{1-\tilde{a}^{2S+1}}. \tag{4.14}$$

Plugging (4.14) into (4.12), we have:

$$\overline{\pi}_{\bar{s}(t)} = \frac{1-\tilde{a}}{1-\tilde{a}^{2S+1}}\tilde{a}^{S+\bar{s}(t)}.$$

Hence Theorem 4.2 holds. ∎

Lemma 4.1 *When $0 < \bar{s}(t) < S$, an AT-STPLA's $q(t)$ is located in state set*

$$\mathscr{A}_{\bar{s}(t)} = \left\{ \frac{\left(1 + \bar{s}(t)\right)\bar{s}(t)}{2} + 1, \dots, 2N - 1, 2N \right\}. \tag{4.15}$$

When $\bar{s}(t) = S$, $q(t)$ is located at

$$\mathscr{A}_{\bar{s}(t)} = \{2N\}. \tag{4.16}$$

Proof: According to AT-STPLA's initialization principle, when $\bar{s}(t) = 0$, $q(t) \in \mathscr{A}_0 = \{1, \dots, 2N\}$.

1) When $\bar{s}(t) = 1$, every state in $\mathscr{A}_0 = \{1, \ldots, 2N\}$ moves to right by one state, such that state 1 cannot be reached and $q(t)$ belongs to $\mathscr{A}_1 = \{2, \ldots, 2N\}$. When $\bar{s}(t) = 2$, every state in $\mathscr{A}_1 = \{2, \ldots, 2N\}$ moves to right by two states, such that states 1, 2, and 3 cannot be reached and $q(t)$ belongs to $\mathscr{A}_2 = \{4, \ldots, 2N\}$.

2) Thus in general, if $0 < \bar{s}(t) = k < S$, an AT-STPLA's current state $q(t)$ must be located in state set $\mathscr{A}_k = \{(1+k)k/2 + 1, \ldots, 2N - 1, 2N\}$.

3) If $0 < \bar{s}(t) = k + 1 < S$, elements in $\mathscr{A}_{k+1}$ are obtained by which the elements in $\mathscr{A}_k$ move to right by $k+1$ states. Therefore, $\mathscr{A}_{k+1} = \{(1+k)k/2 + 1 + (k+1), \ldots, 2N\} = \{(k+2)(k+1)/2 + 1, \ldots, 2N\}$.

Thus when $0 < \bar{s}(t) < S = \left\lceil \left(\sqrt{16N-7} - 1\right)/2 \right\rceil$,

$$\mathscr{A}_{\bar{s}(t)} = \left\{ \frac{(1+\bar{s}(t))\,\bar{s}(t)}{2} + 1, \ldots, 2N - 1, 2N \right\}.$$

When $\bar{s}(t) = S = \left\lceil \left(\sqrt{16N-7} - 1\right)/2 \right\rceil$, every state in $\mathscr{A}_{S-1} = \{S(S-1)/2 + 1, \ldots, 2N-1, 2N\}$ moves to right by $(S-1)$ states. Because $S(S-1)/2 + 1 + (S-1) = (S+1)S/2 + 1 \geq 2N$, it holds that $\mathscr{A}_S = \{2N\}$, according to the state transition principle of AT-STPLA. ■

For example when $N = 5$ in Fig. 4.4, once $\bar{s}(t)$ is fixed, $\mathscr{A}_{\bar{s}(t)}$ is shown intuitively.

Theorem 4.3 *If $\bar{p} > 0.5$ ($\tilde{a} > 1$), when $N \to \infty$, the Alert probability $\mathbb{P}_1 \to 0$.*

Proof: Given any integer $\mathbb{Z}$, with $-S < \mathbb{Z} \leq S$, one can divide the whole space of step states $\{-S, \ldots, S\}$ into two parts: $\{-S, \ldots, \mathbb{Z} - 1\}$ and $\{\mathbb{Z}, \ldots, S\}$. According to the Theorem of Total Probability [13], unfolding $\mathbb{P}_2$, we obtain

$$
\begin{aligned}
\mathbb{P}_2 &= Pr\left(q(t) \in \mathbb{S}_2\right) \\
&= Pr\left(q(t) \in \mathbb{S}_2 \middle| -S \leq \bar{s}(t) \leq \mathbb{Z} - 1\right) \\
&\quad \times Pr\left(-S \leq \bar{s}(t) \leq \mathbb{Z} - 1\right) \\
&\quad + Pr\left(q(t) \in \mathbb{S}_2 \middle| \mathbb{Z} \leq \bar{s}(t) \leq S\right) \\
&\quad \times Pr\left(\mathbb{Z} \leq \bar{s}(t) \leq S\right).
\end{aligned}
\tag{4.17}
$$

Now we assign a value to $\mathbb{Z}$ as follows

$$\mathbb{Z} = \left\lceil \frac{\sqrt{8N+1} - 1}{2} \right\rceil > 0,
\tag{4.18}$$

and plug it into (4.17), thus yielding:

$$
\begin{aligned}
\mathbb{P}_2 &= Pr\left(q(t) \in \mathbb{S}_2\right) \\
&= Pr\left(q(t) \in \mathbb{S}_2 \Big| -S \leq \bar{s}(t) \leq \left\lceil \tfrac{\sqrt{8N+1}-1}{2} \right\rceil - 1\right) \\
&\quad \times Pr\left(-S \leq \bar{s}(t) \leq \left\lceil \tfrac{\sqrt{8N+1}-1}{2} \right\rceil - 1\right) \\
&\quad + Pr\left(q(t) \in \mathbb{S}_2 \Big| \left\lceil \tfrac{\sqrt{8N+1}-1}{2} \right\rceil \leq \bar{s}(t) \leq S\right) \\
&\quad \times Pr\left(\left\lceil \tfrac{\sqrt{8N+1}-1}{2} \right\rceil \leq \bar{s}(t) \leq S\right).
\end{aligned}
\tag{4.19}
$$

In (4.19), we first compute the result of the probability

$$
Pr\left(q(t) \in \mathbb{S}_2 \Big| \left\lceil \frac{\sqrt{8N+1}-1}{2} \right\rceil \leq \bar{s}(t) \leq S\right),
\tag{4.20}
$$

which means that for $\mathbb{Z} \leq \bar{s}(t) \leq S$, $q(t)$ belongs to Pattern Tracking States $\{N+1, \ldots, 2N\}$.

According to Lemma 4.1, when $\mathbb{Z} \leq \bar{s}(t) \leq S$, $q(t)$ is located in state set $\mathscr{A}_{\mathbb{Z} \leq \bar{s}(t) \leq S}$, where

$$
\begin{aligned}
\mathscr{A}_{\mathbb{Z} \leq \bar{s}(t) \leq S} &= \mathscr{A}_{\mathbb{Z}} \cup \cdots \cup \mathscr{A}_{S-1} \cup \mathscr{A}_S \\
&= \{(1+\mathbb{Z})\mathbb{Z}/2 + 1, \ldots, 2N-1, 2N\} \cup \cdots \cup \\
&\quad \{S(S-1)/2 + 1, \ldots, 2N-1, 2N\} \cup \\
&\quad \{(1+S)S/2 + 1, \ldots, 2N-1, 2N\} \\
&= \{(1+\mathbb{Z})\mathbb{Z}/2 + 1, \ldots, 2N-1, 2N\}.
\end{aligned}
\tag{4.21}
$$

Plugging (4.18) into (4.21), we obtain

$$
\begin{aligned}
\mathscr{A}_{\mathbb{Z} \leq \bar{s}(t) \leq S} &= \{N+1, \ldots, 2N-1, 2N\} \\
&= \mathbb{S}_2.
\end{aligned}
\tag{4.22}
$$

Hence, when $\mathbb{Z} \leq \bar{s}(t) \leq S$, $q(t)$ is located in state set $\mathbb{S}_2$, and we can conclude

$$
\begin{aligned}
&Pr\left(q(t) \in \mathbb{S}_2 \Big| \mathbb{Z} \leq \bar{s}(t) \leq S\right) \\
&= Pr\left(q(t) \in \mathscr{A}_{\mathbb{Z} \leq \bar{s}(t) \leq S} \Big| \mathbb{Z} \leq \bar{s}(t) \leq S\right) \\
&= 1.
\end{aligned}
\tag{4.23}
$$

Then, we compute the result of:

$$
Pr\left(\mathbb{Z} = \left\lceil \frac{\sqrt{8N+1}-1}{2} \right\rceil \leq \bar{s}(t) \leq S\right),
\tag{4.24}
$$

in (4.19), where

$$
\begin{aligned}
Pr &\left(\mathbb{Z} = \left\lceil \frac{\sqrt{8N+1}-1}{2} \right\rceil \le \bar{s}(t) \le S \right) \\
&= \sum_{\bar{s}(t)=\mathbb{Z}}^{S} Pr\left(\bar{s}(t)\right) \\
&= \sum_{\bar{s}(t)=\mathbb{Z}}^{S} \overline{\pi}_{\bar{s}(t)} \\
&= \overline{\pi}_{\mathbb{Z}} + \overline{\pi}_{\mathbb{Z}+1} + \cdots + \overline{\pi}_{S}.
\end{aligned}
\tag{4.25}
$$

Plugging (4.12) and (4.14) into (4.25), we obtain

$$
\begin{aligned}
& \overline{\pi}_{\mathbb{Z}} + \overline{\pi}_{\mathbb{Z}+1} + \cdots + \overline{\pi}_{S} \\
&= \left(\tilde{a}^{S+\mathbb{Z}} + \tilde{a}^{S+\mathbb{Z}+1} + \cdots + \tilde{a}^{2S} \right) \overline{\pi}_{-S} \\
&= \tilde{a}^{S+\mathbb{Z}} \frac{1-\tilde{a}^{S-\mathbb{Z}+1}}{1-\tilde{a}} \frac{1-\tilde{a}}{1-\tilde{a}^{2S+1}} \\
&= \frac{\tilde{a}^{S+\mathbb{Z}}-\tilde{a}^{2S+1}}{1-\tilde{a}^{2S+1}} \\
&= \frac{1-\tilde{a}^{\mathbb{Z}-S-1}}{1-\tilde{a}^{-2S-1}} \\
&= \frac{1-\tilde{a}^{\left\lceil \frac{\sqrt{8N+1}-1}{2} \right\rceil - \left\lceil \frac{\sqrt{16N-7}-1}{2} \right\rceil - 1}}{1-\tilde{a}^{-2\left\lceil \frac{\sqrt{16N-7}-1}{2} \right\rceil - 1}}.
\end{aligned}
\tag{4.26}
$$

Observing (4.26), because $\tilde{a} > 1$, we conclude

$$
\lim_{N \to \infty} Pr\left(\mathbb{Z} = \left\lceil \frac{\sqrt{8N+1}-1}{2} \right\rceil \le \bar{s}(t) \le S \right) = 1.
\tag{4.27}
$$

Then combine the results of (4.23) and (4.27), when $N \to \infty$ and $\tilde{p} > 0.5$, the product of expressions (4.20) and (4.24) tends to 1. Thus, $\mathbb{P}_2$ tends to 1, $\mathbb{P}_1 = 1 - \mathbb{P}_2$ tends to 0.

Hence the Theorem 4.3. ∎

Lemma 4.2 *When $-S < \bar{s}(t) < 0$, an AT-STPLA must be located in state set*

$$
q(t) \in \mathscr{A}_{\bar{s}(t)} = \left\{ 1, 2, \ldots, 2N - \frac{\left(\bar{s}(t)-1\right)\bar{s}(t)}{2} \right\}.
\tag{4.28}
$$

When $\bar{s}(t) = -S$, $q(t)$ is located at

$$
\mathscr{A}_{\bar{s}(t)} = \{1\}.
\tag{4.29}
$$

Proof: The proof is similar to the proof of Lemma 4.1. For conciseness, the procedure is omitted. ∎

Theorem 4.4 *If $\tilde{p} < 0.5$ ($0 < \tilde{a} < 1$), when $N \to \infty$, the Alert probability $\mathbb{P}_1 \to 1$.*

Proof: The proof is similar to the proof of Theorem 4.3, which is sketched as follows. Let

$$\mathbb{Z} = \left\lfloor \frac{1 - \sqrt{8N+1}}{2} \right\rfloor < 0. \tag{4.30}$$

Then,

$$
\begin{aligned}
\mathbb{P}_1 &= Pr\left(q(t) \in \mathbb{S}_1\right) \\
&= Pr\left(q(t) \in \mathbb{S}_1 \middle| -S \le \bar{s}(t) \le \left\lfloor \tfrac{1-\sqrt{8N+1}}{2} \right\rfloor \right) \\
&\quad \times Pr\left(-S \le \bar{s}(t) \le \left\lfloor \tfrac{1-\sqrt{8N+1}}{2} \right\rfloor \right) \\
&\quad + Pr\left(q(t) \in \mathbb{S}_1 \middle| \left\lfloor \tfrac{1-\sqrt{8N+1}}{2} \right\rfloor + 1 \le \bar{s}(t) \le S \right) \\
&\quad \times Pr\left(\left\lfloor \tfrac{1-\sqrt{8N+1}}{2} \right\rfloor + 1 \le \bar{s}(t) \le S \right).
\end{aligned}
\tag{4.31}
$$

According to Lemma 4.2, when $\mathbb{Z}$ satisfies (4.30),

$$\mathscr{A}_{-S \le \bar{s}(t) \le \mathbb{Z}} = \mathbb{S}_1. \tag{4.32}$$

Therefore, in (4.31),

$$Pr\left(q(t) \in \mathbb{S}_1 \middle| -S \le \bar{s}(t) \le \mathbb{Z} = \left\lfloor \frac{1-\sqrt{8N+1}}{2} \right\rfloor \right) = 1. \tag{4.33}$$

$$
\begin{aligned}
&Pr\left(-S \le \bar{s}(t) \le \mathbb{Z} = \left\lfloor \tfrac{1-\sqrt{8N+1}}{2} \right\rfloor \right) \\
&= \sum_{\bar{s}(t)=-S}^{\mathbb{Z}} \bar{\pi}_{\bar{s}(t)} \\
&= \frac{1-\tilde{a}^{S+\mathbb{Z}+1}}{1-\tilde{a}^{2S+1}} \\
&= \frac{1-\tilde{a}^{\left\lceil \frac{\sqrt{16N-7}-1}{2} \right\rceil + \left\lfloor \frac{1-\sqrt{8N+1}}{2} \right\rfloor +1}}{2^{\left\lceil \frac{\sqrt{16N-7}-1}{2} \right\rceil +1}}.
\end{aligned}
\tag{4.34}
$$

Thus, considering $0 < \tilde{a} < 1$, we obtain

$$\lim_{N \to \infty} Pr\left(-S \le \bar{s}(t) \le \mathbb{Z} = \left\lfloor \frac{1-\sqrt{8N+1}}{2} \right\rfloor \right) = 1. \tag{4.35}$$

Combining the results of (4.33) and (4.34), we conclude that when $N \to \infty$ and $\tilde{p} < 0.5$, the product of (4.33) and (4.34) approaches 1. Thus, $\mathbb{P}_1$ approaches 1. Hence Theorem 4.4 holds. ∎

Here we introduce a tuning parameter h ($-1 < h < 1$) to express its bias orientation and degree. Note that when $h > 0$, it refers to the bias toward Suppress in a similar way in [8, 9], while $h < 0$ means the bias toward Notify (Alert), and naturally, $h = 0$ means Balanced. Thus, the update principle of $\bar{s}(t)$ is summarized as follows.

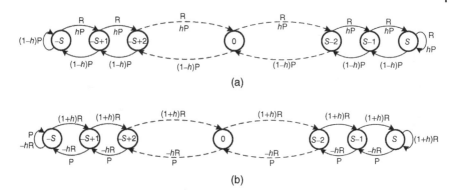

Figure 4.5 The state transition principle of $\bar{s}(t)$ with parameter h. R means *Reward* and P means *Penalty*. Due to the hypothesis of an AT-STPLA, an occurrence of an event corresponding to a *Reward* has probability $\bar{p}$. The nonoccurrence of an event corresponding to a *Penalty* has probability $\bar{q}$. (a) Case of $0 < h < 1$ and (b) Case of $-1 < h < 0$.

1) If $rand() \geq |h|$ or $h = 0$, when *Reward*, update $\bar{s}(t)$ obeying (4.1), and when *Penalty*, obeying (4.2).
2) If $rand() < |h|$ and $h > 0$, update $\bar{s}(t)$ obeying (4.1).
3) If $rand() < |h|$ and $h < 0$, update $\bar{s}(t)$ obeying (4.2).

The step state transition with parameter h is shown in Fig. 4.5.
Then, update $q(t)$ via (4.3).

4.1.5 Simulation Studies

Three series of experimental results are given in this section to demonstrate the excellent capabilities of AT-STPLA in both accuracy and speed during the whole process of pattern discovering and tracking. In the following experiments, it is compared with STPLA and STP-TFSLA under different challenging settings such that the advantages of AT-STPLA can be observed intuitively. We set $N_1 = 12$, $N_2 = 11$, and $N = 12$, respectively, as done in [8].

In this series of experiments, we attempt to highlight AT-STPLA's accuracy in a static environment with omission noise. Hence, we record the Alert probabilities of STP-TFSLA and AT-STPLA, respectively. All the settings are summarized as follows and the simulation results are drawn in Fig. 4.6.

1) $\mathbb{P}_1$ reported here is obtained by repeating the experiment 10 000 times, each time consisting of 5000 iterations.
2) Omission noise probability $\bar{q}$ varies from 0 to 0.3.
3) The initial state $q(0)$ of STP-TFSLA and AT-STPLA is all chosen randomly.

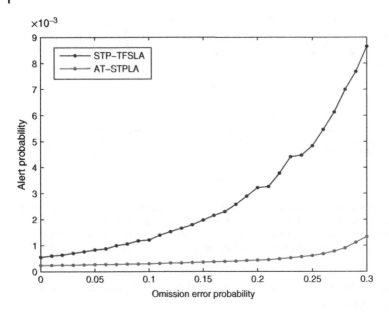

Figure 4.6 The mean alert probability under different omission error probabilities when $N_1 = N_2 + 1 = N = 12$.

4) h of STP-TFSLA equals 0.2 and h of AT-STPLA equals 0. Actually, it is a comparison between Suppress-biased STP-TFSLA and balanced AT-STPLA, which is more challenging and "unfair" to AT-STPLA.

Experimental results in Fig. 4.6 demonstrate that in a static environment with omission noise, AT-STPLA outperforms STP-TFSLA in accuracy. It is not hard to deduce that AT-STPLA is also more accurate than STPLA, because STPLA is STP-TFSLA when $h = 0$.

In this series of experiments, we attempt to highlight the AT-STPLA's accuracy under a static environment with inclusion noise. Hence, we record the Alert probabilities of STP-TFSLA and AT-STPLA, respectively. All the settings are summarized as follows, and the simulation results are drawn in Fig. 4.7.

1) $\mathbb{P}_1$ reported here is obtained by repeating the experiment 10 000 times, each time consists of 5000 iterations.
2) Inclusion noise probability $\tilde{p}$ varies from 0 to 0.3.
3) The initial state $q(0)$ of STP-TFSLA and AT-STPLA is chosen randomly.
4) $h = 0$ in STP-TFSLA and AT-STPLA. Because STP-TFSLA is not able to provide an Alert bias and in this condition it degenerates to STPLA.

Experimental results in Fig. 4.7 demonstrate that in a static environment with inclusion noise, AT-STPLA outperforms STP-TFSLA and STPLA in accuracy.

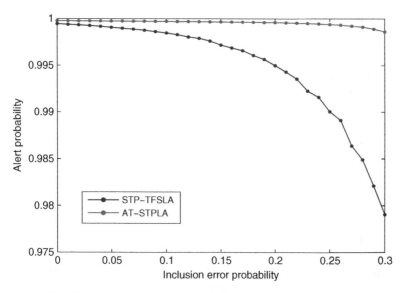

Figure 4.7 The mean alert probability under different inclusion error probabilities when $N_1 = N_2 + 1 = N = 12$.

In this series of experiments, we arrange AT-STPLA, STP-TFSLA, and STPLA to perform in various dynamic environments in order to compare their pattern discovering and tracking speed. AT-STPLA with different h values is taken into account to show the preference ability toward either Suppress or Notify (Alert), which its peers lack. The results are shown in Figs. 4.9–4.14, involving three kinds of noisy environments with different values of $\tilde{p}$ and $\tilde{q}$. Note that the final convergence curves are obtained after 10 000 simulations, and every single simulation contains 600 iterations. The pattern changes every 150 iterations, $q(0)$ is chosen randomly, and $N_1 = N_2 + 1 = N = 12$, i.e., the same configuration as that in [8].

From Figs. 4.8–4.13, we draw two conclusions: (i) When h values of the STP-TFSLA and AT-STPLA are equal, the speed of AT-STPLA is much faster than that of STP-TFSLA. AT-STPLA establishes a significant advantage in just 150 iterations; (ii) For AT-STPLA, as shown in Figs. 4.8–4.10, when $h > 0$, the Suppress-biased performance is getting more and more inclined with h's increase. With the increasing $\tilde{p}$ or $\tilde{q}$, such bias is getting more and more distinct. While in Figs. 4.11–4.13, when $h < 0$, the Alert-biased performance is getting more and more inclined as h decreases. As $\tilde{p}$ or $\tilde{q}$ increases, such bias is getting more and more distinct as well.

From the above extensive experimental results, we conclude that the introduced scheme not only satisfies the user requirements for accuracy and convergence rate but also significantly outperforms the existing online learning mechanisms in terms of learning speed.

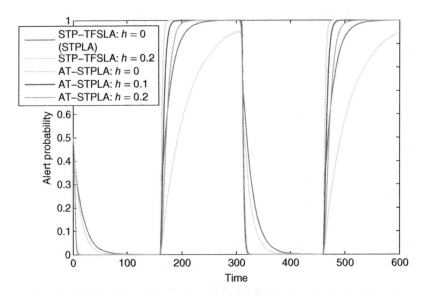

Figure 4.8 Respective performances of AT-STPLA and STP-TFSLA adopting different Suppress-biased tune parameters, when omission noise $\tilde{q} = 0.1$ or inclusion noise $\tilde{p} = 0.1$.

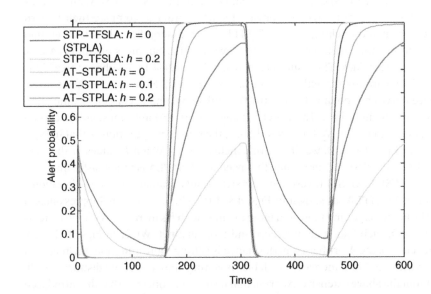

Figure 4.9 Respective performances of AT-STPLA and STP-TFSLA adopting different Suppress-biased tune parameters, when omission noise $\tilde{q} = 0.2$ or inclusion noise $\tilde{p} = 0.2$.

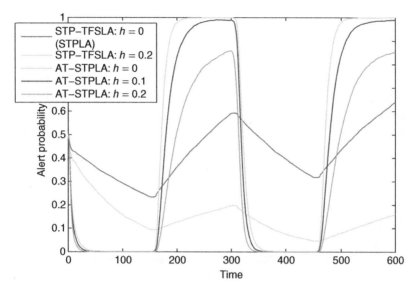

Figure 4.10 Respective performances of AT-STPLA and STP-TFSLA adopting different Suppress-biased tune parameters, when omission noise $\tilde{q} = 0.3$ or inclusion noise $\tilde{p} = 0.3$.

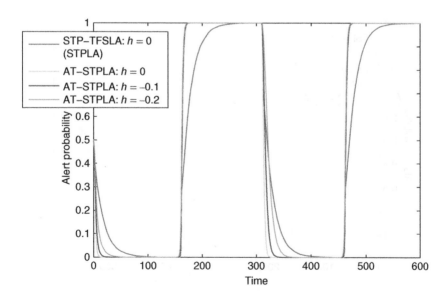

Figure 4.11 Respective performances of AT-STPLA and STP-TFSLA adopting different Alert-biased tune parameters, when omission noise $\tilde{q} = 0.1$ or inclusion noise $\tilde{p} = 0.1$.

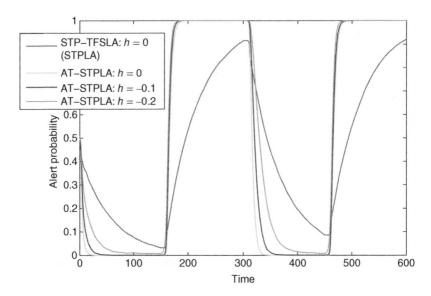

Figure 4.12 Respective performances of AT-STPLA and STP-TFSLA adopting different Alert-biased tune parameters, when omission noise $\tilde{q} = 0.2$ or inclusion noise $\tilde{p} = 0.2$.

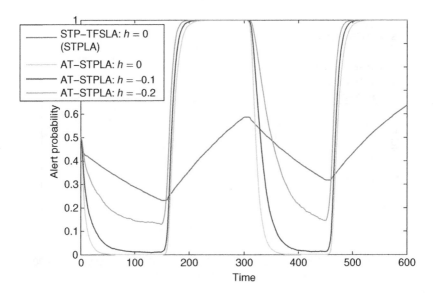

Figure 4.13 Respective performances of AT-STPLA and STP-TFSLA adopting different Alert-biased tune parameters, when omission noise $\tilde{q} = 0.3$ or inclusion noise $\tilde{p} = 0.3$.

4.1.6 Summary

Each LA's performance becomes a key in the routine pattern learning framework [5]. This section first introduces a Variable Structure Stochastic Automaton into the field of spatiotemporal pattern discovery and tracking to enhance the LAs in [5, 8]. Its variable state transition generates a more accurate and faster convergence, both of which are pressing requirements for spatiotemporal pattern learning in real-world applications. Theoretical analyses are given, and three experiments are performed to demonstrate the introduced LA's excellent learning efficiency and tunable bias preference, even if the noise reaches the 30% level. Spatiotemporal pattern learning automata have essential applications in real life, which are applied to various fields.

4.2 Stochastic Searching on the Line

A Stochastic Point Location (SPL) problem is another application-oriented learning automaton that aims to find a target parameter on a one-dimensional line by operating a controlled random walk and receiving information from a Stochastic EnVironment (SEV). If the target parameter changes randomly, we call the parameter *dynamic*; otherwise *static*. SEV mainly consists of two types of environments – *informative* ($\hat{p} > 0.5$ where $\hat{p}$ represents the probability for an environment providing a correct suggestion) and *deceptive* ($\hat{p} < 0.5$). Hierarchical Stochastic Searching on the Line (HSSL) is the most efficient algorithm to catch static or dynamic parameters in an informative environment. However, they are unable to locate the target parameter in a deceptive environment and unable to recognize an environment's type (informative or deceptive). This section presents a Symmetrical HSSL (SHSSL) [14] by extending an HSSL binary tree-based search structure to a symmetrical form. By means of this innovative way, the introduced learning mechanism can converge to a static or dynamic target parameter in the range of not only $0.618 < \hat{p} < 1$, but also $0 < \hat{p} < 0.382$. It is noted that 0.618 is an approximate value of the Golden Ratio conjugate [15]. In [16] Yazidi *et al.* demonstrate that HSSL's effective range must be greater than the value of Golden Ratio, and they use 0.618 to substitute the value of Golden Ratio. Hereinafter, we also use quantity 0.618 to denote the conjugate of the Golden Ratio. The experimental results show that this scheme is efficient and feasible to solve the SPL problem in both informative and deceptive environments.

4.2.1 Background and Motivation

A point location problem can be solved via a Learning Mechanism (LM) by moving on a one-dimensional domain (for example, the unity interval) and attempting to locate a particular point. An environment ("Oracle") communicates with it to assist the LM and suggests which direction it should move to (left or right). Once the LM received such a suggestion, it would move to the next point according to the suggestion. At the new point, LM continues to communicate with SEV to obtain a new suggestion. In this way, the procedure is repeated till some termination criteria are met.

If the information provided by an environment is deterministic, it is a Deterministic Point Location (DPL) problem that has been investigated thoroughly. Correspondingly, if the environment suggests the LM to move to a correct direction with the probability of $\hat{p}$ ($0 < \hat{p} < 1$) while to a wrong direction with the probability of $(1 - \hat{p})$, namely, if the environment is stochastic, it becomes a SPL problem. SPL is pioneered by Oommen [17, 18]. Generally speaking, SEV could also be divided into two types [18], i.e., informative and deceptive. In the former, SEV tells the truth, guiding the LM to a correct direction with $0.5 < \hat{p} < 1$. While in the latter, SEV tells the truth with $0 < \hat{p} < 0.5$ to mislead the LM in the wrong direction.

Moreover, Oommen *et al.* [19] arrange SPL in a more difficult condition, i.e., the non-stationary environment. In SPL, the most specific measure of an environment is the probability $\hat{p}$. If $\hat{p}$ is constant in the whole search process, we name this environment a stationary one; otherwise, if $\hat{p}$ changes randomly at any possible time instant, it is called a non-stationary environment, which means an environment can even transform between a stochastic teacher and a stochastic compulsive liar. In addition, as a search aim, the target parameter λ^* itself may change as well. Hence, broadly speaking, a stationary environment should possess a fixed $\hat{p}$ and a fixed λ^*. If not, the environment is said to be non-stationary.

In SPL-related investigations, Learning Automaton (LA) [20–23] is a robust implementation that can be used in online and offline learning. With the aid of LAs' convergence process, SPL algorithms can approach the target parameter. SPL, like the fundamental LA problem, is of importance in its own right and also on the merit of the potential that it has in all LA-based applications [16].

SPL has been found to be applied in many potential or direct ways including [16]: power management in smart grids [24], distributed channel selection [25], solving the minimum weight connected dominating set [26], classification [27], power control [28], service selection [29], solving a large class of wireless-networks-related problems [30], a general class of stochastic decentralized games [31], adaptive control of antennas in wireless push networks [32], and in optimal sensor placement [33]. In [34], Oommen *et al.* show how a Multidimensional Scaling (MDS) scheme can be enhanced by incorporating into it an SPL strategy, which optimizes the former's gradient descent learning phase and calls the new algorithm as MDS-SPL. In [35], Oommen and Calitoiu construct a model at the first phase of training and then utilize the SPL scheme in [17] to learn an unknown key-value – contagion parameter at the second phase of prediction. Nevertheless, if the learning model is inaccurate and deceptive, a high-efficiency deceptive SPL algorithm is required further if an application scenario in [24–31, 33] accords with an SPL model. However, we do not know the environment's type (informative or deceptive, stationary or non-stationary). SHSSL introduced in this section is then needed to be applied because we can deal with this problem without knowing an environment's type.

Currently, many studies have been done for solving SPL. We summarize their introduced methods as follows.

Stochastic Search on the Line (SSL): To solve SPL, Oommen [17] first introduces a scheme for searching the optimal parameter in an informative environment. In SSL, the unity interval is discretized into $\hat{N}$ subintervals at the position of $\{0, \frac{1}{\hat{N}}, \frac{2}{\hat{N}}, \ldots, \frac{\hat{N}-1}{\hat{N}}, 1\}$, and a larger $\hat{N}$ will lead a higher accuracy. LM receives SEV's suggestion and obeys it absolutely.

Continuous Point Location with Adaptive Tertiary Search (CPL-ATS): Oommen and Raghunath [36] introduce CPL-ATS, which resorts LAs to determine a valid interval. It mainly consists of two steps, construction of LAs and interval elimination. The current interval is partitioned into three disjoint subintervals in every search epoch. After LAs' learning process, an interval containing no target parameter is eliminated. By repeating these two steps, the current interval containing the target parameter is getting more accurate.

Continuous Point Location with Adaptive $\bar{d}$-ARY Search (CPL-AdS): Instead of CPL-ATS, CPL-AdS [18] is presented by enlarging the number of LAs from three to $\bar{d}$. It extends the initial unity interval to $[-\frac{1}{\bar{d}-2}, 1 + \frac{1}{\bar{d}-2})$, and an environment type can be judged in the first search epoch – if an environment is deceptive, there is no decision table entry for the decision vector $\vec{\Omega}$, which leads the next search interval to $[0, 1)$ and inverts penalties and rewards. Compared with CPL-ATS, CPL-AdS has the advantage of a flexible number of LAs, but the decision table needs to be generated manually depending on the value of $\bar{d}$.

General CPL-AdS: As an extension, Huang and Jiang [37] generalize the CPL-AdS scheme. The scheme inherits all the merits of CPL-AdS. Furthermore, the decision formula substitutes decision tables and the extending formula assists to fix varying parameters efficiently and automatically. These two techniques make General CPL-AdS the only scheme catching both static and dynamic parameters in informative and deceptive environments so far.

Adaptive Step Searching (ASS): Tao *et al.* [38] utilize the historical information of past three decisions to determine the next search step size. In this case, if the historical information tells LM that the target parameter is far away from the current point, LM should then magnify the step size in order to approach the target quickly. If historical information enlightens LM that the target parameter is nearby, LM should thus diminish the step size to catch the parameter subtly. So far, ASS is the fastest algorithm for solving SPL, but its incapability in a deceptive environment and low stability are two deficiencies.

Hierarchical Stochastic Searching on the Line (HSSL): Yazidi *et al.* arrange SPL into a binary search tree in [16]. The scheme possesses fast convergence and high accuracy and is indeed the most efficient algorithm for solving SPL currently when $0.618 < \hat{p} < 1$. However, it is unable to locate the target parameter in a deceptive environment and unable to recognize an environment's type (informative or deceptive).

Here we give a summary table to intuitively illustrate the SPL-related algorithms' capability under different types of environments, as shown in Table 4.1. Note that the existing algorithms for solving a deceptive SPL are based on learning automata (LAs). If an LM attempts to judge the SEV's type, extra computing resources and search time are needed.

In this section, we introduce a novel scheme extending HSSL's search structure to a symmetrical one to deal with SPL in both informative and deceptive environments. The Symmetrical HSSL (SHSSL) retains all the advantages of a hierarchical searching scheme in an informative environment and possesses the same capacity in a deceptive environment. Therefore, it can locate static or dynamic parameters in both informative and deceptive environments when $0.618 < \hat{p} < 1$ and $0 < \hat{p} < (1 - 0.618) = 0.382$ by fast speed and high accuracy, without any additional computing resource and search time, whether the environment is stationary or not.

In the original HSSL, the entire search process is arranged in a form of a full binary tree with its root at the top and depth $\overline{D} = log_2(\hat{N})$, where $\hat{N}$ is the resolution of the algorithm. For example, when $\overline{D} = 3$, the binary tree is shown in Fig. 4.14. As a binary tree grows, $\overline{D}$ becomes larger (i.e., tree's depth becomes larger),

Table 4.1 Capability under different environments of all SPL algorithms.

Environments Algorithms	Informative	Deceptive	Stationary	Non-stationary
SSL	√	×	√	√
CPL-ATS	√	×	√	×
CPL-AdS	√	√	√	×
General CPL-AdS	√	√	√	√
ASS	√	×	√	√
HSSL	√	×	√	√
SHSSL	√	√	√	√

Note that a stationary environment consists of stationary $\hat{p}$ and static λ^*.

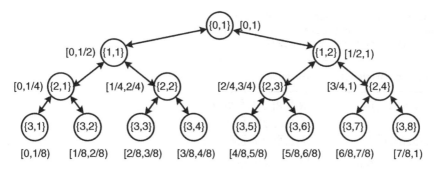

Figure 4.14 Original HSSL's search structure is a binary tree. The search space when $\overline{D} = 3$. Source: [16]/with permission of IEEE.

which leads to a higher resolution. Notation $\overline{\Delta}_{\{\overline{d},\overline{j}\}}$, represents the search interval $[(\overline{j} - 1)(\frac{1}{2})^{\overline{d}}, \overline{j}(\frac{1}{2})^{\overline{d}}]$, where $\overline{d}$ indicates the current depth and $\overline{j}$ indicates the sequential index of a node in the $\overline{d}$-th layer. For example, $\overline{\Delta}_{\{3,5\}}$ indicates interval $[\frac{4}{8}, \frac{5}{8})$.

In each single search instant, HSSL partitions the current search interval $\overline{\Delta}_{\{\overline{d},\overline{j}\}}$ into two disjoint subintervals whose sizes are equal. Automatically, they become the left and right child intervals, respectively.

At any given instant, LM may arrive at the interval $\overline{\Delta}_{\{\overline{d},\overline{j}\}}$, where $\forall \overline{j} \in \{1, 2, \ldots, 2^{\overline{d}}\}$ and $0 \leq \overline{d} \leq \overline{D}$, and visit three points in $\overline{\Delta}_{\{\overline{d},\overline{j}\}}$ noted as $\overline{x}_1$, $\overline{x}_2$, and $\overline{x}_3$, generating a 3-tuple $\vec{x} = [\overline{x}_1, \overline{x}_2, \overline{x}_3]$, where $\overline{x}_1 = \overline{\sigma}_{\{\overline{d},\overline{j}\}} = (\overline{j} - 1)(\frac{1}{2})^{\overline{d}}$, $\overline{x}_2 = mid(\overline{\Delta}_{\{\overline{d},\overline{j}\}}) = (2\overline{j} - 1)(\frac{1}{2})^{\overline{d}+1}$ and $\overline{x}_3 = \overline{\gamma}_{\{\overline{d},\overline{j}\}} = \overline{j}(\frac{1}{2})^{\overline{d}}$. $\overline{x}_1, \overline{x}_2$, and $\overline{x}_3$ equal the left bound, midpoint, and right bound of $\overline{\Delta}_{\{\overline{d},\overline{j}\}}$, respectively. Meanwhile, LM adopts $\overline{x}_2 = mid(\overline{\Delta}_{\{\overline{d},\overline{j}\}})$ as the current estimate of the true parameter value λ^*. After sampling three points of $\vec{x} = [\overline{x}_1, \overline{x}_2, \overline{x}_3]$, it then communicates with SEV to get a response tuple noted as $\vec{\Omega} = [\Omega_1, \Omega_2, \Omega_3]$, in which, correspondingly, Ω_k is the $\overline{x}_k$'s relative direction to λ^*, for $k \in \{1, 2, 3\}$. Note that SEV obeys the principle [16] as follows to generate a decision tuple $\vec{\Omega}$.

If $\lambda^* < x^k$

$$\Omega_k = \begin{cases} \vec{L}, & \text{with probability } \hat{p} \\ \vec{R}, & \text{with probability } (1 - \hat{p}). \end{cases} \tag{4.36}$$

If $\lambda^* \geq x^k$

$$\Omega_k = \begin{cases} \vec{L}, & \text{with probability } (1 - \hat{p}) \\ \vec{R}, & \text{with probability } \hat{p}. \end{cases} \tag{4.37}$$

Table 4.2 HSSL decision table and SHSSL informative table to choose next search interval in different conditions of tuple $\vec{\Omega}$ [16].

Next search interval	Condition
$Parent(\overline{\Delta}_{\{i,j\}})$	$[\vec{R}, \vec{R}, \vec{R}] \bigvee [\vec{L}, \vec{R}, \vec{R}] \bigvee$ $[\vec{L}, \vec{L}, \vec{R}] \bigvee [\vec{L}, \vec{L}, \vec{L}]$
$LeftChild(\overline{\Delta}_{\{i,j\}})$	$[\vec{R}, \vec{L}, \vec{L}] \bigvee [\vec{R}, \vec{L}, \vec{R}]$
$RightChild(\overline{\Delta}_{\{i,j\}})$	$[\vec{R}, \vec{R}, \vec{L}] \bigvee [\vec{L}, \vec{R}, \vec{L}]$

a) $\overline{\Delta}_{\{i,j\}}$ denotes the current search interval.

Thus, as to LM, 2^3 possible responses, i.e., $[\vec{L}, \vec{L}, \vec{L}]$, $[\vec{L}, \vec{L}, \vec{R}]$, $[\vec{L}, \vec{R}, \vec{L}]$, $[\vec{L}, \vec{R}, \vec{R}]$, $[\vec{R}, \vec{L}, \vec{L}]$, $[\vec{R}, \vec{L}, \vec{R}]$, $[\vec{R}, \vec{R}, \vec{L}]$, and $[\vec{R}, \vec{R}, \vec{R}]$, can be received. Based on $\vec{\Omega}$, the next search interval is chosen by referring to the HSSL decision table illustrated in Table 4.2.

Resorting Kelly's theory [39], any tree structure associated with a finite stationary Markov process is time-reversible. Moreover, the scheme of HSSL is affirmed to be asymptotically optimal if $\hat{p}$ is larger than the conjugate of the Golden Ratio by analyzing the properties of the underlying Markov chain [16]. In other words, HSSL has its own limitation, i.e., it converges only when $0.618 < \hat{p} < 1$. The pseudo-code of HSSL is stated in Algorithm HSSL.

HSSL: Hierarchical Stochastic Searching on the Line Algorithm

Input: Tree depth $\bar{D}$, environment coefficient $\hat{p}$
Output: $\tilde{\lambda}(t)$: estimate of λ^* at time t
 function CALNEXTINTERVAL(CurrentInterval, $\vec{\Omega}$)
  result $\leftarrow$ Invoke Table 4.2
  return *result*
 end function
 CurrentInterval $\leftarrow \bar{\Delta}_{\{0,1\}}$
 for TimeInstant $= 1 \rightarrow t$ **do**
  $\tilde{\lambda}(t) \leftarrow mid$(CurrentInterval)
  $\vec{x} \leftarrow [\bar{\sigma}, mid, \bar{\gamma}]$
  $\vec{\Omega} \leftarrow$ Feedback of SEV at $\vec{x}$
  CurrentInterval $\leftarrow$ CalNextInterval(CurrentInterval, $\vec{\Omega}$)
 end for
 END

Example 4.4 We denote Ω^* to be the correct response obtained from a nonfaulty environment (i.e., an environment for which $\hat{p} = 1$). We now consider the case displayed in the example illustrated in Fig. 4.14, where node $\overline{S}_{\{3,7\}}$ is the target node. Thus, in this particular example, $\lambda^* \in [6/8, 7/8)$. Consider now the transitions at node $\overline{S}_{\{1,2\}}$. Its associated interval, $[1/2, 1)$, is sampled at the points $\vec{x} = [1/2, 3/4, 1]$. Taking into account that $\lambda^* \in [6/8, 7/8)$, one can easily see that $\Omega^* = [\vec{R}, \vec{R}, \vec{L}]$.

4.2.2 Symmetrical Hierarchical Stochastic Searching on the Line

The search structure of SHSSL is also a complete binary tree inheriting from the original HSSL scheme, but it is in the form of a symmetrical structure as shown in Fig. 4.15.

For convenience, we use the same notations as those in [40] and [16]. Because of the symmetry of SHSSL, we define a symmetrical tree's depth $\overline{D}$ as the depth of a lower or upper half tree, where $\overline{D} > 0$, distinctly. A larger $\hat{N}$ leads to higher accuracy. In SHSSL, $\hat{N} = 2^{\overline{D}}$. The lower half of a symmetrical tree. As a matter of fact, an informative tree in SHSSL whose layer count $\overline{d}$ is from 0 to $\overline{D}$ is identical to the binary tree in HSSL. As shown in Fig. 4.2, the root node of an informative

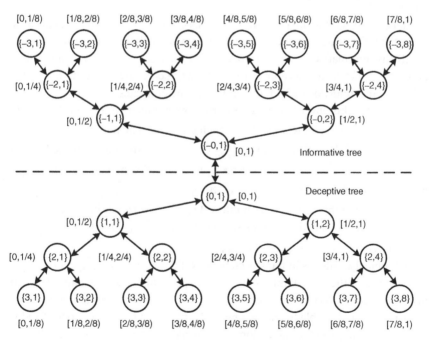

Figure 4.15 Symmetrical HSSL's search structure. The search space when $\overline{D} = 3$.

tree (informative root node) is at layer 0, and the leaf nodes of an informative tree are at layer $\bar{D}$ in the lower half of the entire tree.

The upper half of a symmetrical tree. What is different from an informative tree is that the layer count in a deceptive tree is set as a negative integer such that LM is able to recognize in which tree itself lies. As shown in Fig. 4.15, the root node of a deceptive tree (deceptive root node) is at layer -0 and the leaf nodes of a deceptive tree are at layer $(-\bar{D})$.

$\bar{S}_{\{\bar{d},\bar{j}\}}$ is defined as the $\bar{j}$-th ($\forall \bar{j} \in \{1, 2, \ldots, 2^{\bar{d}}\}$) tree node at layer $\bar{d}$, while $\bar{S}_{\{-\bar{d},\bar{j}\}}$ is the $\bar{j}$-th ($\forall \bar{j} \in \{1, 2, \ldots, 2^{\bar{d}}\}$) tree node at layer $-\bar{d}$, where $0 \leq \bar{d} \leq \bar{D}$. Therefore, some correspondence come into being. Specifically, node $\bar{S}_{\{\bar{d},\bar{j}\}}$'s corresponding node in the upper half of the tree is $\bar{S}_{\{-\bar{d},\bar{j}\}}$, and both of them represent the interval $[(\bar{j} - 1)(\frac{1}{2})^{\bar{d}}, \bar{j}(\frac{1}{2})^{\bar{d}})$. Particularly, $\bar{S}_{\{0,1\}}$ and $\bar{S}_{\{-0,1\}}$ are the root nodes and $\bar{S}_{\{\bar{D},\bar{j}\}}$ and $\bar{S}_{\{-\bar{D},\bar{j}\}}$ are the leaf nodes where $\forall \bar{j} \in \{1, 2, \ldots, 2^{\bar{D}}\}$.

It is defined as the search interval. $\bar{\sigma} = (\bar{j} - 1)(\frac{1}{2})^{\bar{d}}$ is the left bound and $\bar{\gamma} = \bar{j}(\frac{1}{2})^{\bar{d}}$ is the right bound. An interval $\bar{\Delta}_{\{\bar{d},\bar{j}\}}$ corresponds to an informative tree's node $\bar{S}_{\{\bar{d},\bar{j}\}}$ and an interval $\bar{\Delta}_{\{-\bar{d},\bar{j}\}}$ corresponds to a deceptive tree's node $\bar{S}_{\{-\bar{d},\bar{j}\}}$. Note that intervals $\bar{\Delta}_{\{\bar{d},\bar{j}\}}$ and $\bar{\Delta}_{\{-\bar{d},\bar{j}\}}$ denote the exactly same interval in spite of their different forms.

A target node is at depth $\bar{D}$ or $-\bar{D}$ and is a special one among leaf nodes because it contains the target parameter λ^*, which is what SHSSL attempts to locate. If an SEV is informative, the target node lies in the informative tree, while if it is deceptive, it lies in the deceptive tree. A target node's corresponding interval is called a target interval at which the target parameter locates.

At each time instant, LM partitions current search interval $\bar{\Delta}_{\{\bar{d},\bar{j}\}}$ or $\bar{\Delta}_{\{-\bar{d},\bar{j}\}}$ into two equally sized subintervals at the midpoint. The left portion $\bar{\Delta}_{\{\bar{d}+1,2\bar{j}-1\}}$ or $\bar{\Delta}_{\{-(\bar{d}+1),2\bar{j}-1\}}$ and the right portion $\bar{\Delta}_{\{\bar{d}+1,2\bar{j}\}}$ or $\bar{\Delta}_{\{-(\bar{d}+1),2\bar{j}\}}$ are defined as the left and right children of $\bar{\Delta}_{\{\bar{d},\bar{j}\}}$ or $\bar{\Delta}_{\{-\bar{d},\bar{j}\}}$, respectively. Inversely, interval $\bar{\Delta}_{\{\bar{d},\bar{j}\}}$ or $\bar{\Delta}_{\{-\bar{d},\bar{j}\}}$ is the parent of intervals $\bar{\Delta}_{\{\bar{d}+1,2\bar{j}-1\}}$ or $\bar{\Delta}_{\{-(\bar{d}+1),2\bar{j}-1\}}$ and $\bar{\Delta}_{\{\bar{d}+1,2\bar{j}\}}$ or $\bar{\Delta}_{\{-(\bar{d}+1),2\bar{j}\}}$. Especially, when the current search interval is $\bar{\Delta}_{\{0,1\}}$, its parent interval is $\bar{\Delta}_{\{-0,1\}}$, and vice versa. When the current search depth is $\bar{D}$ or $-\bar{D}$, $\bar{\Delta}_{\{\bar{D},\bar{j}\}}$ or $\bar{\Delta}_{\{-\bar{D},\bar{j}\}}$'s children, $\forall \bar{j} \in \{1, 2, \ldots, 2^{\bar{D}}\}$, are themselves.

Undoubtedly, SHSSL's initial search node is still the root node ($\bar{S}_{\{0,1\}}$ or $\bar{S}_{\{-0,1\}}$). But because of the uncertainty of an SEV's type, before a search starts, LM can make a random selection between $\bar{S}_{\{0,1\}}$ and $\bar{S}_{\{-0,1\}}$ to be its initial search node. Alternatively, LM is able to choose one of the two introduced root nodes fixedly. These two strategies of initialization are equivalent because both of them have a 50% chance to start in a correct tree (i.e., if the environment is informative, a search starts at $\bar{S}_{\{0,1\}}$; if deceptive, the search starts at $\bar{S}_{\{-0,1\}}$). If the initial search node is chosen incorrectly, the SHSSL algorithm can still lead LM to converge, and it has no effect on the convergence result.

As same as HSSL, at any given time instant, LM could arrive at interval $\overline{\Delta}_{\{\overline{d},j\}}$ or $\overline{\Delta}_{\{-\overline{d},j\}}$, where $\forall j \in \{1, 2, \ldots, 2^{\overline{d}}\}$ and $0 \leq \overline{d} \leq \overline{D}$, and visit three points in $\overline{\Delta}_{\{\overline{d},j\}}$ or $\overline{\Delta}_{\{-\overline{d},j\}}$ noted as $\overline{x}_1$, $\overline{x}_2$, and $\overline{x}_3$, generating a 3-tuple $\vec{x} = [\overline{x}_1, \overline{x}_2, \overline{x}_3]$, where $\overline{x}_1 = \overline{\sigma}_{\{\overline{d},j\}} = (j-1)(\frac{1}{2})^{\overline{d}}$, $\overline{x}_2 = mid(\overline{\Delta}_{\{\overline{d},j\}}) = (2j-1)(\frac{1}{2})^{\overline{d}+1}$, and $\overline{x}_3 = \overline{\gamma}_{\{\overline{d},j\}} = j(\frac{1}{2})^{\overline{d}}$, respectively. Meanwhile, LM adopts $\overline{x}_2$ of $\vec{x}$ to be the current estimate of target parameter λ^* as done in [16]. Especially, if the algorithm makes LM converge at the target node (interval), $\overline{x}_2$ of the target node (interval) is the algorithm's correctly converged value.

After SEV visiting $\vec{x} = [\overline{x}_1, \overline{x}_2, \overline{x}_3]$ at $\overline{S}_{\{\overline{d},j\}}$ or $\overline{S}_{\{-\overline{d},j\}}$, it generates a suggestion about λ^*'s relative location to $\overline{x}_k$ ($\vec{L}$ or $\vec{R}$), for $k \in \{1, 2, 3\}$. Therefore, three suggestions are generated simultaneously and we note these three suggestions as Ω_1, Ω_2, and Ω_3, forming a decision vector $\vec{\Omega} = [\Omega_1, \Omega_2, \Omega_3]$. For $k \in \{1, 2, 3\}$, if $\lambda^* < x^k$, SEV generates Ω_k obeying (1), while if $\lambda^* \geq x^k$, Ω_k is determined via (2).

Thus, eight possible responses, i.e., $[\vec{L}, \vec{L}, \vec{L}]$, $[\vec{L}, \vec{L}, \vec{R}]$, $[\vec{L}, \vec{R}, \vec{L}]$, $[\vec{L}, \vec{R}, \vec{R}]$, $[\vec{R}, \vec{L}, \vec{L}]$, $[\vec{R}, \vec{L}, \vec{R}]$, $[\vec{R}, \vec{R}, \vec{L}]$, and $[\vec{R}, \vec{R}, \vec{R}]$, are generated. Based on $\vec{\Omega}$, if the current search node lies in an informative tree ($\overline{S}_{\{\overline{d},j\}}$), the next search interval is chosen by referring to the informative table illustrated in Table 4.2. Accordingly, if the current search node lies in a deceptive tree ($\overline{S}_{\{-\overline{d},j\}}$), LM uses Table 4.3 to decide the next search interval.

Applying (1)–(2) and above decision tables, LM can jump into a new search node (interval), in which LM can obtain a new round of $\vec{\Omega}$ at $\vec{x}$ to jump into another new search node (interval). Repeating this procedure until a given search time is reached, the value of λ^* can be estimated, and naturally, the SEV's type can be judged by the type of the tree at which LM converges – if converging at an informative tree, SEV's type is informative, while if converging at a deceptive tree, it is deceptive.

The pseudo-code of SHSSL is given as Algorithm SHSSL.

Table 4.3 The deceptive table of SHSSL to choose next search interval in different conditions of tuple $\vec{\Omega}$.

Next search interval	Condition
$Parent(\overline{\Delta}_{\{i,j\}})$	$[\vec{R}, \vec{R}, \vec{R}] \bigvee [\vec{R}, \vec{R}, \vec{L}] \bigvee$ $[\vec{R}, \vec{L}, \vec{L}] \bigvee [\vec{L}, \vec{L}, \vec{L}]$
$LeftChild(\overline{\Delta}_{\{i,j\}})$	$[\vec{L}, \vec{R}, \vec{R}] \bigvee [\vec{L}, \vec{R}, \vec{L}]$
$RightChild(\overline{\Delta}_{\{i,j\}})$	$[\vec{L}, \vec{L}, \vec{R}] \bigvee [\vec{R}, \vec{L}, \vec{R}]$

SHSSL: Symmetrical Hierarchical Stochastic Searching on the Line Algorithm

Input: Tree depth $\bar{D}$, environment coefficient $\hat{p}$
Output: $\tilde{\lambda}(t)$: estimate of λ^* at time t
    function CALNEXTINTERVAL(CurrentInterval, $\vec{\Omega}$)
        if the CurrentInterval is in informative tree **then**
            result ← Invoke Table 4.2
        else
            result ← Invoke Table 4.3
        end if
        return *result*
    end function
    CurrentInterval ← $\bar{\Delta}_{\{0,1\}}$ or $\bar{\Delta}_{\{-0,1\}}$
    for TimeInstant = 1 → t **do**
        $\lambda(t)$ ← mid(CurrentInterval)
        $\vec{x}$ ← $[\bar{\sigma}, mid, \bar{\gamma}]$
        $\vec{\Omega}$ ← Feedback of SEV at $\vec{x}$
        CurrentInterval ← CalNextInterval(CurrentInterval, $\vec{\Omega}$)
    end for
    END

Example 4.5 In order to elaborate on how our scheme catches a target parameter, we give a trace of execution of an example for the case of $\bar{D} = 3$, $\hat{p} = 0.2$ (a deceptive environment). The target parameter is set as 0.9123, which is a benchmark value used in [16, 17, 41], and the initial search node is set as $\bar{S}_{\{0,1\}}$ of the informative tree.

Step 1: Current Interval: $\bar{\Delta}_{\{0,1\}} = [0, 1)$
         $\vec{x} = [0, 0.5, 1]$
         $\vec{\Omega} = [\vec{R}, \vec{L}, \vec{L}]$
         Next Interval: $\bar{\Delta}_{\{1,1\}} = [0, 0.5)$
Step 2: Current Interval: $\bar{\Delta}_{\{1,1\}} = [0, 0.5)$
         $\vec{x} = [0, 0.25, 0.5]$
         $\vec{\Omega} = [\vec{L}, \vec{L}, \vec{L}]$
         Next Interval: $\bar{\Delta}_{\{0,1\}} = [0, 1)$
Step 3: Current Interval: $\bar{\Delta}_{\{0,1\}} = [0, 1)$
         $\vec{x} = [0, 0.5, 1]$
         $\vec{\Omega} = [\vec{L}, \vec{L}, \vec{R}]$
         Next Interval: $\bar{\Delta}_{\{-0,1\}} = [0, 1)$

Step 4: Current Interval: $\overline{\Delta}_{\{-0,1\}} = [0, 1)$
$\vec{x} = [0, 0.5, 1]$
$\vec{\Omega} = [\vec{L}, \vec{L}, \vec{R}]$
Next Interval: $\overline{\Delta}_{\{-1,2\}} = [0.5, 1)$

Step 5: Current Interval: $\overline{\Delta}_{\{-1,2\}} = [0.5, 1)$
$\vec{x} = [0.5, 0.75, 1]$
$\vec{\Omega} = [\vec{L}, \vec{R}, \vec{R}]$
Next Interval: $\overline{\Delta}_{\{-2,3\}} = [0.5, 0.75)$

Step 6: Current Interval: $\overline{\Delta}_{\{-2,3\}} = [0.5, 0.75)$
$\vec{x} = [0.5, 0.625, 0.75]$
$\vec{\Omega} = [\vec{L}, \vec{L}, \vec{L}]$
Next Interval: $\overline{\Delta}_{\{-1,2\}} = [0.5, 1)$

Step 7: Current Interval: $\overline{\Delta}_{\{-1,2\}} = [0.5, 1)$
$\vec{x} = [0.5, 0.75, 1]$
$\vec{\Omega} = [\vec{L}, \vec{L}, \vec{R}]$
Next Interval: $\overline{\Delta}_{\{-2,4\}} = [0.75, 1)$

Step 8: Current Interval: $\overline{\Delta}_{\{-2,4\}} = [0.75, 1)$
$\vec{x} = [0.75, 0.875, 1]$
$\vec{\Omega} = [\vec{R}, \vec{L}, \vec{R}]$
Next Interval: $\overline{\Delta}_{\{-3,8\}} = [0.875, 1)$

Step 9: Current Interval: $\overline{\Delta}_{\{-3,8\}} = [0.875, 1)$
$\vec{x} = [0.875, 0.9375, 1]$
$\vec{\Omega} = [\vec{L}, \vec{R}, \vec{R}]$
Next Interval: $\overline{\Delta}_{\{-3,8\}} = [0.875, 1)$

The algorithm converges at target interval $\overline{\Delta}_{\{-3,8\}} = [0.875, 1)$ successfully, and LM adopts $\frac{0.875+1}{2} = 0.9375$ as the estimate of the target parameter 0.9123 at Step 9. Moreover, SEV's type is deceptive obviously because LM can note that the final convergence node is in the deceptive tree.

Obviously, the introduced symmetrical structure is not only suitable for HSSL but also fits all discretized and adaptive SPL schemes, including SSL and ASS, for solving deceptive SPL. Especially after Jiang *et al.* introduce a triple-level SPL [42], symmetrizing a searching structure is expected to be an efficient methodology to overcome the restriction in its so-called unstable region. Symmetrizing a search structure is indeed a powerful thought that can be used to solve deceptive SPL without any extra computing resources and search time.

4.2.3 Simulation Studies

In this section, we present three series of representative experimental results to extensively demonstrate SHSSL's capacity in a deceptive environment when

$0 < \hat{p} < 0.382$ and its efficiency compared with HSSL and ASS in an informative environment when $0.618 < \hat{p} < 1$, even if $\hat{p}$ is changing with time.

In the first series of experiments, we record SHSSL and HSSL's respective mean estimate values for various values of $\hat{p}$ and tree depth $\overline{D} = log_2(\hat{N})$ when λ^* is fixed at 0.9123 (a conventional target parameter applied in experiments of [41]). Each mean estimate value is obtained after performing SHSSL for 10^7 iterations, i.e., the time instant t is as large as 10^7. This quantity of 10^7 was also adopted in the prior work [41]. The results are given in Table 4.4, in which we add HSSL's mean

Table 4.4 SHSSL and HSSL's mean estimate value under different conditions of various $\hat{p}$ and $\overline{D}$ when $\lambda^* = 0.9123$.

SHSSL	$\hat{p} = 0.95$	$\hat{p} = 0.85$	$\hat{p} = 0.70$	$\hat{p} = 0.30$	$\hat{p} = 0.15$	$\hat{p} = 0.05$
$\overline{D} = 2$	0.8666097	0.8298404	0.6828313	0.6825750	0.8296846	0.8665891
$\overline{D} = 3$	0.9332183	0.9111990	0.7669322	0.7662220	0.9110511	0.9332360
$\overline{D} = 4$	0.9079078	0.9087951	0.8192048	0.8185083	0.9086654	0.9079099
$\overline{D} = 5$	0.9210162	0.9183753	0.8550496	0.8555317	0.9183583	0.9210178
$\overline{D} = 6$	0.9144895	0.9153531	0.8778202	0.8778897	0.9153219	0.9144902
$\overline{D} = 7$	0.9104170	0.9114768	0.8905981	0.8902931	0.9114799	0.9104180
$\overline{D} = 8$	0.9120050	0.9118859	0.8988607	0.8987180	0.9118846	0.9120048
$\overline{D} = 9$	0.9130206	0.9127750	0.9039892	0.9044332	0.9127755	0.9130206
$\overline{D} = 10$	0.9126237	0.9126574	0.9073851	0.9071918	0.9126577	0.9126237
$\overline{D} = 11$	0.9123698	0.9124323	0.9093575	0.9092731	0.9124321	0.9123698
$\overline{D} = 12$	0.9122397	0.9122809	0.9104909	0.9106290	0.9122812	0.9122397

HSSL	$\hat{p} = 0.95$	$\hat{p} = 0.85$	$\hat{p} = 0.70$	$\hat{p} = 0.30$	$\hat{p} = 0.15$	$\hat{p} = 0.05$
$\overline{D} = 2$	0.8666203	0.8309458	0.7037266	0.4384165	0.4604388	0.4869599
$\overline{D} = 3$	0.9332186	0.9114942	0.7844115	0.4254259	0.4573474	0.4868118
$\overline{D} = 4$	0.9079084	0.9088419	0.8318240	0.4177364	0.4562901	0.4868093
$\overline{D} = 5$	0.9210158	0.9183867	0.8638002	0.4126746	0.4559586	0.4868440
$\overline{D} = 6$	0.9144894	0.9153566	0.8835128	0.4094955	0.4558356	0.4868169
$\overline{D} = 7$	0.9104167	0.9114775	0.8942233	0.4076170	0.4558247	0.4868375
$\overline{D} = 8$	0.9120050	0.9118860	0.9012334	0.4059050	0.4556079	0.4868139
$\overline{D} = 9$	0.9130205	0.9127751	0.9054470	0.4049817	0.4557548	0.4868465
$\overline{D} = 10$	0.9126237	0.9126575	0.9083362	0.4034745	0.4557077	0.4868295
$\overline{D} = 11$	0.9123698	0.9124322	0.9098059	0.4035725	0.4556315	0.4868208
$\overline{D} = 12$	0.9122397	0.9122809	0.9108321	0.4034042	0.4557251	0.4868174

estimate value to compare the convergence property of these two schemes. Note that in this series of experiments, we set the initial search node as $\overline{S}_{\{0,1\}}$ of the informative tree uniformly because in [41] the initial node is the root node, and each result is rounded and retained with 7 decimal numbers after the decimal point.

We can conclude from Table 4.4 that in both informative and deceptive environments, it is observable that whatever $\hat{p}$ and $\overline{D}$ are fixed at, SHSSL's mean estimate value always approaches the true value of λ^*, which means that just a small error exists in the results by executing the introduced scheme. With the increase of $\overline{D}$, the mean estimate value is getting more accurate.

In the informative environment, the two schemes have a similar performance in accuracy as expected. In the deceptive aspect, compared with HSSL, SHSSL works efficiently and the same as the circumstance in the informative environment, while HSSL attempts to converge at the root node, which is a totally wrong direction. With the decrease of $\hat{p}$ ($\hat{p} < 0.5$), SHSSL's advantage in convergence and accuracy is remarkable. Hence, this series of experiments show that SHSSL and HSSL perform analogously in an informative environment, but SHSSL has a massive advantage over HSSL in a deceptive environment, in which HSSL loses its ability to converge.

In the second series of experiments, to present the convergence speed and capacity in both informative and deceptive environments of our scheme visibly, we execute SHSSL, HSSL, and ASS (the fastest SPL algorithm) to catch a dynamic parameter in a parallel way. Meanwhile, $\hat{p}$ is fixed in its effective interval for SHSSL (i.e., $\hat{p} = 0.05$, $\hat{p} = 0.15$, $\hat{p} = 0.30$, $\hat{p} = 0.70$, $\hat{p} = 0.85$, and $\hat{p} = 0.95$) to contrast the performance of the three schemes. In each simulation, 1000 parallel experiments are conducted in order to generate an accurate ensemble average of results, following [16]. Note that λ^* in this series of experiments varies periodically between 0.9123 and $(1 - 0.9123)$, and $\overline{D}$ is fixed to be 10, which means resolution $\hat{N} = 1024$. In this series of experiments, the initial search node is still set as $\overline{S}_{\{0,1\}}$.

In Fig. 4.16a, λ^* changes every 100 iterations and $\overline{D} = 10$, $\hat{N}_{max} = 1024$. Three algorithms run simultaneously in the informative environment when $\hat{p} = 0.95$. We can observe that in the whole process, three algorithms are almost synchronous. But according to Fig. 4.16b, where λ^* switches every 100 iterations and SEV is deceptive when $\hat{p} = 0.05$, we can see an entirely different circumstance from Fig. 4.16a—it shows that SHSSL converges the same as the condition in Fig. 4.16a, while HSSL and ASS fail to do so.

With the decrease of $\hat{p}$, the convergence speed of the three algorithms is getting slower and slower according to Fig. 4.17a where $\overline{D} = 10$, $\hat{N}_{max} = 1024$. The period increases to 400 iterations, and $\hat{p}$ is fixed at 0.85. But the three algorithms still keep almost the same speed and accuracy. While in Fig. 4.17a where the target parameter switches every 400 iterations and $\hat{p}$ is as low as 0.15 (deceptive), SHSSL

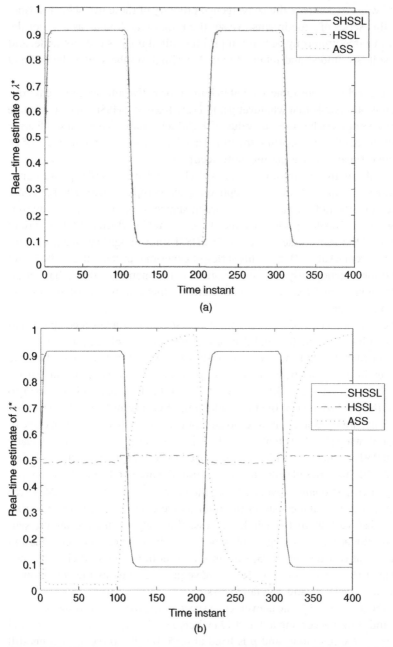

Figure 4.16 The learning characteristics under opposite value of $\hat{p}$ (0.95 and 0.05).
(a) Case where λ^* switches between 0.9123 and $(1 - 0.9123)$ every 100 iterations,
$\hat{p} = 0.95$ and (b) Case where λ^* switches between 0.9123 and $(1 - 0.9123)$ every 100
iterations, $\hat{p} = 0.05$.

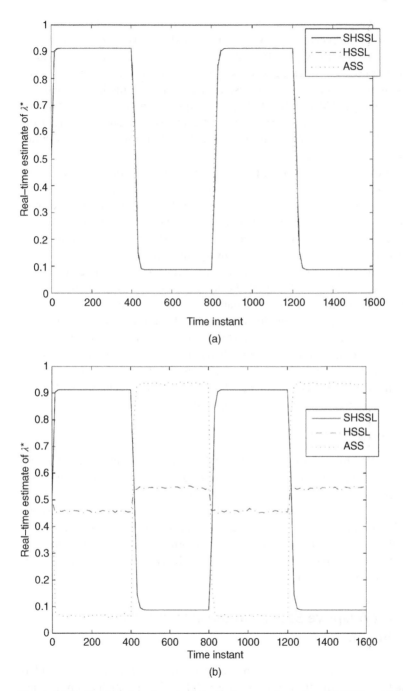

Figure 4.17 The learning characteristics under opposite value of $\hat{p}$ (0.85 and 0.15). (a) Case where λ^* switches between 0.9123 and $(1 - 0.9123)$ every 400 iterations, $\hat{p} = 0.85$ and (b) Case where λ^* switches between 0.9123 and $(1 - 0.9123)$ every 400 iterations, $\hat{p} = 0.15$.

retains its stable performance to catch the dynamic parameters, but HSSL and ASS are deceived by the deceptive environment so as to lose its convergence direction.

In Fig. 4.18a, $\overline{D} = 10$, $\hat{N}_{max} = 1024$, $\hat{p} = 0.70$, and λ^* switches every 1000 iterations. On this occasion, three algorithms are still largely identical, but they slacken their speed and the stability is decreasing as well because the lower value of $\hat{p}$ brings ambiguity to LM such that LM's jump frequency among nodes (SHSSL and HSSL) or points (ASS) is increasing. In the deceptive circumstance in Fig. 4.18b where $\hat{p} = 0.30$, HSSL and ASS malfunction, while SHSSL functions well as shown in Fig. 4.18a, which is not surprising.

So from the second series of experiments (Figs. 4.16–4.18), we can conclude when $\overline{D} = 10$, in an informative environment ($\hat{p} = 0.95$, $\hat{p} = 0.85$, and $\hat{p} = 0.70$), SHSSL performs as well as HSSL in terms of speed and accuracy. In a deceptive environment, SHSSL has a huge advantage over HSSL and ASS because SHSSL's convergence property has no difference in both informative and deceptive environments, while HSSL and ASS fail to converge in a deceptive environment.

In the third series of experiments, target parameter λ^* continues to switch between 0.9123 and $(1 - 0.9123)$ periodically, and $\overline{D}$ equals ten as before. Because of the dynamic nature of a non-stationary SEV, $\hat{p}$ can be valued randomly with time. Here we consider a more difficult case, in which p is picked from not only $0.618 < p < 1$ but also $0 < \hat{p} < 0.382$. That is, $\hat{p}$'s value is stochastic in its valid scope at every time instant. The probability for $\hat{p}$ to take values in $(0.618, 1)$ $\hat{p}_{informative}$ is following [42]. Figures 4.19–4.21 demonstrate the capability of SHSSL under a non-stationary environment where both $\hat{p}$ and λ^* change with time. Clearly, the introduced method works very well.

4.2.4 Summary

In order to perform stochastic Searching on the Line well in both informative and deceptive environments, this section extends the binary search tree into a symmetrical structure and thus introduces Symmetrical Hierarchical Stochastic Searching on the Line (SHSSL). Three series of experiments are performed to prove visually that SHSSL is efficient in informative and deceptive environments. Thus, it can catch both static and dynamic parameters in informative and deceptive environments when $0.618 < \hat{p} < 1$ and $0 < \hat{p} < 0.382$, with fast speed and high accuracy, whether $\hat{p}$ is stationary or not.

4.3 Fast Adaptive Search on the Line in Dual Environments

The other non-hierarchical method named Adaptive Step Search (ASS) [38] has been the fastest algorithm so far for solving a SPL problem, which can be applied

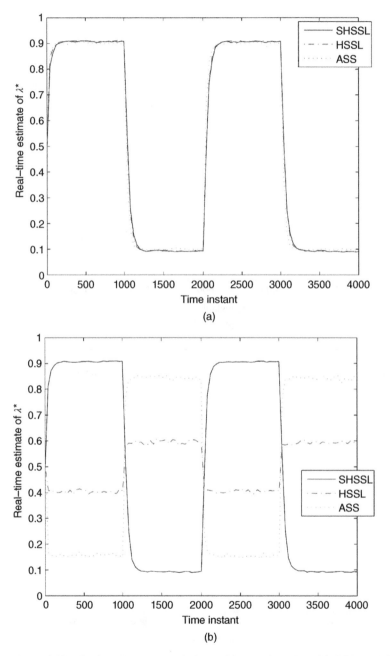

Figure 4.18 The learning characteristics under opposite value of $\hat{p}$ (0.70 and 0.30). (a) Case where λ^* switches between 0.9123 and $(1 - 0.9123)$ every 1000 iterations, $\hat{p} = 0.70$ and (b) Case where λ^* switches between 0.9123 and $(1 - 0.9123)$ every 1000 iterations, $\hat{p} = 0.30$.

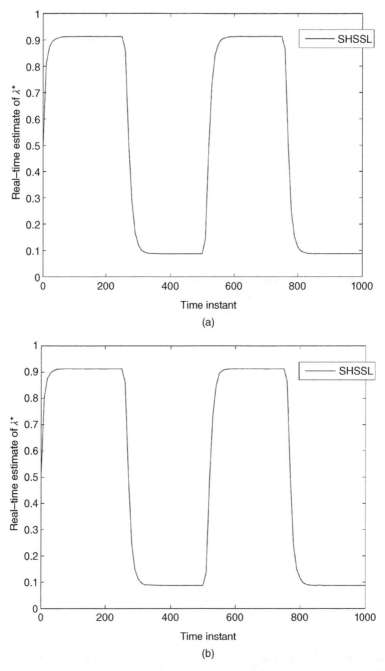

Figure 4.19 The learning characteristics under opposite value of $\hat{p}_{informative}$ (0.9 and 0.1). (a) Case where λ^* switches between 0.9123 and $(1 - 0.9123)$ every 250 iterations, $\hat{p}_{informative} = 0.9$ and (b) Case where λ^* switches between 0.9123 and $(1 - 0.9123)$ every 250 iterations, $\hat{p}_{informative} = 0.1$.

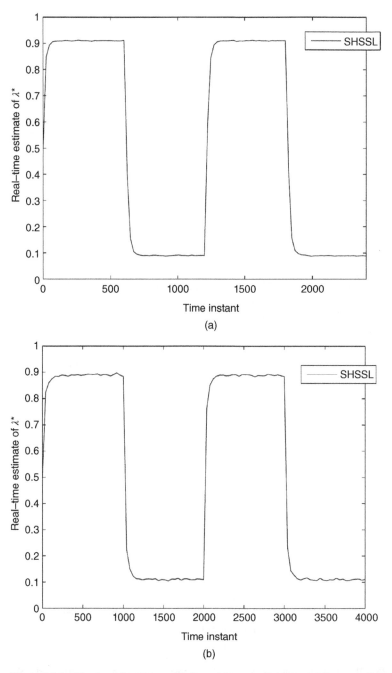

Figure 4.20 The learning characteristics under opposite value of $\hat{p}_{informative}$ (0.8 and 0.2). (a) Case where λ^* switches between 0.9123 and $(1 - 0.9123)$ every 600 iterations, $\hat{p}_{informative} = 0.8$ and (b) Case where λ^* switches between 0.9123 and $(1 - 0.9123)$ every 600 iterations, $\hat{p}_{informative} = 0.2$.

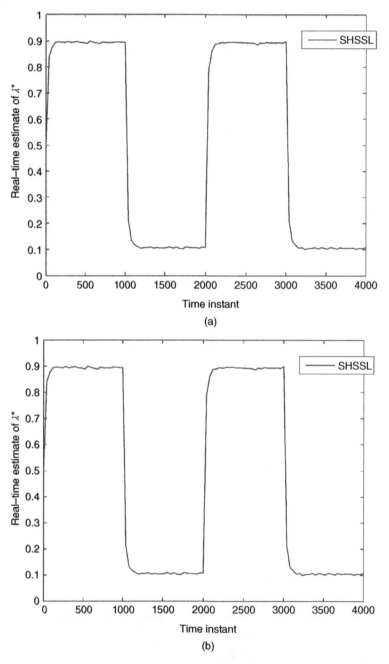

Figure 4.21 The learning characteristics under opposite value of $\hat{p}_{informative}$ (0.7 and 0.3). (a) Case where λ^* switches between 0.9123 and $(1 - 0.9123)$ every 1000 iterations, $\hat{p}_{informative} = 0.7$ and (b) Case where λ^* switches between 0.9123 and $(1 - 0.9123)$ every 1000 iterations, $\hat{p}_{informative} = 0.3$.

to Particle Swarm Optimization (PSO), the establishment of epidemic models, and many other scenarios. However, its application is theoretically restrained within the range of informative environment in which the probability of an environment providing a correct suggestion is strictly bigger than a half. Namely, it does not work in a deceptive environment where such a probability is less than half.

In this section, we introduce a promotion to overcome the difficult issue facing Adaptive Step Search by means of symmetrization and buffer techniques. The algorithm can operate a controlled random walk in both informative and deceptive environments and eventually converge without performance loss. Experimental results demonstrate that the proposed scheme is efficient and feasible in dual environments.

4.3.1 Background and Motivation

ASS [38] roots from the conventional SPL method by discretizing the interval, and its major contribution is the adaptive step size in an SPL solution process. In ASS, three new definitions are introduced: (i) Let $\hat{N}_{min}$ be the smallest resolution, matching the biggest search step size $1/\hat{N}_{min}$; (ii) Let N_{max} be the largest resolution, matching the smallest search step size $1/\hat{N}_{max}$; (iii) Let $\hat{N}(t)$ be the immediate resolution at time instant t, meaning that the search step size is $1/\hat{N}(t)$. Parameter $\hat{N}_{max}$ determines the accuracy of the final convergence, and parameter $\hat{N}_{min}$ affects the speed of finding a target point.

When the target point is far away from the current point, by adjusting the value of $\hat{N}(t)$, a larger step size is used such that the LM is able to reach the target point faster. Inversely, if the target point is located in the surrounding domain, a smaller step size is adopted to approach the target point prudently. Naturally, at any given instant t, $\hat{N}_{min} \leq \hat{N}(t) \leq \hat{N}_{max}$.

The core idea of ASS is the current step size $\hat{N}(t)$ varying with time, strictly obeying the principle of a decision table as shown in Table 4.5, which guarantees its adaptation and fast convergence. In Table 4.5, every feedback from an environment is represented as 0 or 1, where 0 means left and 1 means right. The 3-bits memory of LM stores the historical information of the last three feedbacks including eight conditions $\{000, 001, 010, 011, 100, 101, 110, 111\}$ as an important clue for the next adjustment. The state transition of these conditions is shown in Fig. 4.22. Synthesizing Fig. 4.22 and Table 4.5, we conclude that when the condition is 000 or 111, representing three continuous identical feedbacks, the target point is considered to be far away from the current point. Therefore, $\hat{N}(t) = \hat{N}(t-1)/2$ and the step size doubles. In another case, when the condition is 101 or 010, representing the last three moves are not consistent, the target point is considered nearby. Thus, $\hat{N}(t) = \hat{N}(t-1) * 2$ and the step size is diminished by half. Note

Table 4.5 Decision table of original ASS as well as SASSB in informative environment.

Condition	Current resolution	Current value
000	$\hat{N}(t) = \hat{N}(t-1)/2$	$\tilde{\lambda}(t) = \tilde{\lambda}(t-1) - 1/\hat{N}(t)$
001	$\hat{N}(t) = \hat{N}(t-1)$	$\tilde{\lambda}(t) = \tilde{\lambda}(t-1) - 1/\hat{N}(t)$
010	$\hat{N}(t) = \hat{N}(t-1) * 2$	$\tilde{\lambda}(t) = \tilde{\lambda}(t-1) - 1/\hat{N}(t)$
011	$\hat{N}(t) = \hat{N}(t-1)$	$\tilde{\lambda}(t) = \tilde{\lambda}(t-1) - 1/\hat{N}(t)$
100	$\hat{N}(t) = \hat{N}(t-1)$	$\tilde{\lambda}(t) = \tilde{\lambda}(t-1) + 1/\hat{N}(t)$
101	$\hat{N}(t) = \hat{N}(t-1) * 2$	$\tilde{\lambda}(t) = \tilde{\lambda}(t-1) + 1/\hat{N}(t)$
110	$\hat{N}(t) = \hat{N}(t-1)$	$\tilde{\lambda}(t) = \tilde{\lambda}(t-1) + 1/\hat{N}(t)$
111	$\hat{N}(t) = \hat{N}(t-1)/2$	$\tilde{\lambda}(t) = \tilde{\lambda}(t-1) + 1/\hat{N}(t)$

Source: Adapted from [38].

that during the whole search procedure, constraint $\hat{N}_{min} \leq \hat{N}(t) \leq \hat{N}_{max}$ should be strictly observed.

The pseudo-code of ASS is given in Algorithm ASS.

Symmetrization is a significant thought first applied in SHSSL [14]. Aiming at the specific binary tree search structure, the symmetrization operation mirrors the original tree and the decision table to enable valid performance in both informative and deceptive environments. This promotion is concise and graceful. It does not devastate the intrinsic tree structure and retains the existing excellent performance meanwhile adding the new ability to an existing method.

Similar to HSSL, ASS can be classified as a stochastic algorithm and relies on a specific search structure to reach the target point. Therefore, the symmetrization technique has a potential application to ASS.

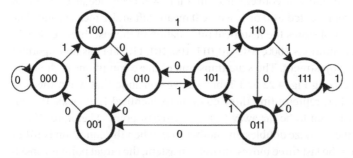

Figure 4.22 Eight possibilities in the 3-bits memory. Source: Adapted from [38].

ASS: Adaptive Step Searching Algorithm

Input: smallest resolution $\hat{N}_{min}$, largest resolution $\hat{N}_{max}$, current resolution $\hat{N}$

Output: $\tilde{\lambda}$: estimate value of target point λ^*

    function CALNEXTRESOLUTION($\hat{N}$)

        result ← Invoke Table 4.2

        return *result*

    end function

    function CALNEXTPOINT($\tilde{\lambda}$)

        result ← Invoke Table 4.2

        return *result*

    end function

    Randomly initialize the 3-bits memory from set $\{000, 001, 010, 011, 100, 101, 110, 111\}$

    $\tilde{\lambda}$ ← 0.5

    for $t = 1 \rightarrow UpperBound$ **do**

        $\hat{N}$ ← CalNextResolution($\hat{N}$)

        $\tilde{\lambda}$ ← CalNextPoint($\tilde{\lambda}$)

    end for

    END

4.3.2 Symmetrized ASS with Buffer

In Symmetrized ASS with Buffer (SASSB), the main notation inherits all the representations in ASS, which are listed as follows.

1) $\hat{N}_{max}$ represents the maximum resolution, matching the minimum step size $1/\hat{N}_{max}$.
2) $\hat{N}_{min}$ represents the minimum resolution, matching the maximum step size $1/\hat{N}_{min}$.
3) $\hat{N}(t)$ represents the current resolution at instant t, matching the step size $1/\hat{N}(t)$ at t. $\hat{N}(t)$ is subject to constraint $\hat{N}_{min} \leq \hat{N}(t) \leq \hat{N}_{max}$.
4) $\tilde{\lambda}(t)$ represents the value of the current point at t.
5) $\bar{b} > 0$ represents the length of the buffer.

As shown in Fig. 4.23, the search structure is redesigned, containing two symmetrical sections. The left one is called the deceptive section, and the right one is the informative one. The buffer is shown as a dotted line segment with a length of $\bar{b}$ in both informative and deceptive sections. Note that the solid line part in the former is identical to that in ASS.

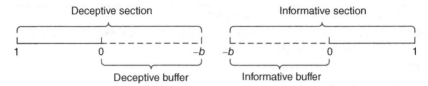

Deceptive section   Informative section

1   0   −b   −b   0   1

Deceptive buffer   Informative buffer

Figure 4.23 Search structure of SASSB.

Generally, the initial search point is located at $-\overline{b}$ in either the informative or deceptive section because the type of environment is not known initially. The initial state determining the step size is randomly chosen from set $\{000, 001, 010, 011, 100, 101, 110, 111\}$.

In any arbitrary instant t, the stored last three feedbacks in the 3-bit memory update according to the existing state transition principle shown in Fig. 4.22. Then, if $\tilde{\lambda}(t-1)$ is located in the informative section, refer to Table 4.5 to determine the current resolution $\hat{N}(t)$, step size $1/\hat{N}(t)$, and location $\tilde{\lambda}(t)$. If $\tilde{\lambda}(t-1)$ is located in the deceptive section, then refer to Table 4.6 to determine the corresponding ones.

When some preset criteria are met, we consider that the type of a section where an LM is located eventually is the type of the SEV.

It is necessary to indicate that in the total search structure, there are two exceptional cases: (i) When $\tilde{\lambda}(t-1)$ is informative 1 (i.e., point 1 in an informative section), if SEV suggests right, $\tilde{\lambda}(t)$ stays unchanged. When $\tilde{\lambda}(t-1)$ is deceptive 1, if SEV suggests left, $\tilde{\lambda}(t)$ stays unchanged either. (ii) When $\tilde{\lambda}(t-1)$ is informative $-\overline{b}$, if SEV suggests left, $\tilde{\lambda}(t)$ then jumps to deceptive $-\overline{b}$. When $\tilde{\lambda}(t-1)$ is deceptive $-\overline{b}$, if SEV suggests right, $\tilde{\lambda}(t)$ then jumps to informative $-\overline{b}$.

The pseudo-code of SASSB is given in Algorithm SASSB.

Table 4.6 The decision table of SASSB in deceptive environment.

Condition	Current resolution	Current value
000	$\hat{N}(t) = \hat{N}(t-1)/2$	$\tilde{\lambda}(t) = \tilde{\lambda}(t-1) + 1/\hat{N}(t)$
001	$\hat{N}(t) = \hat{N}(t-1)$	$\tilde{\lambda}(t) = \tilde{\lambda}(t-1) + 1/\hat{N}(t)$
010	$\hat{N}(t) = \hat{N}(t-1) * 2$	$\tilde{\lambda}(t) = \tilde{\lambda}(t-1) + 1/\hat{N}(t)$
011	$\hat{N}(t) = \hat{N}(t-1)$	$\tilde{\lambda}(t) = \tilde{\lambda}(t-1) + 1/\hat{N}(t)$
100	$\hat{N}(t) = \hat{N}(t-1)$	$\tilde{\lambda}(t) = \tilde{\lambda}(t-1) - 1/\hat{N}(t)$
101	$\hat{N}(t) = \hat{N}(t-1) * 2$	$\tilde{\lambda}(t) = \tilde{\lambda}(t-1) - 1/\hat{N}(t)$
110	$\hat{N}(t) = \hat{N}(t-1)$	$\tilde{\lambda}(t) = \tilde{\lambda}(t-1) - 1/\hat{N}(t)$
111	$\hat{N}(t) = \hat{N}(t-1)/2$	$\tilde{\lambda}(t) = \tilde{\lambda}(t-1) - 1/\hat{N}(t)$

SASSB: Symmetrized Adaptive Step Searching with Buffer Algorithm

Input: minimum resolution $\hat{N}_{min}$, maximum resolution $\hat{N}_{max}$, current resolution $\hat{N}$

Output: $\tilde{\lambda}$: estimate value of target point λ^*

    function CALNEXTRESOLUTION($\hat{N}$, Section)
        if the Section is informative **then**
            result ← Invoke Table 4.5
        else
            result ← Invoke Table 4.6
        end if
        return *result*
    end function
    function CALNEXTPOINT($\tilde{\lambda}$, Section)
        if the Section is informative **then**
            result ← Invoke Table 4.5
        else
            result ← Invoke Table 4.6
        end if
        return *result*
    end function
    Randomly initialize the 3-bits memory from set $\{000, 001, 010, 011, 100, 101, 110, 111\}$
    Section ← informative or deceptive
    $\tilde{\lambda}$ ← $-\bar{b}$ of Section
    for $t = 1 \to UpperBound$ **do**
        $\hat{N}$ ← CalNextResolution($\hat{N}$, Section)
        $\tilde{\lambda}$ ← CalNextPoint($\tilde{\lambda}$, Section)
    end for
    END

Buffer is a pivotal design in SASSB to make full use of the symmetrization technique. We give some analyses related to it.

When $\bar{b} = 0$ or $\bar{b}$ is extremely small, the buffer area does not exist, and the search structure is simplified as Fig. 4.24. Meanwhile, imagine that a target point λ^* is extremely close to 0, and SEV is informative. When the current point is approaching λ^* from the right, because λ^* is near to informative 0, the current point can jump over the deceptive λ^* by mistake with a larger step size and then lead to deceptive 1. Similarly, if SEV is deceptive, the current point can be led to informative 1.

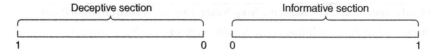

Figure 4.24 Search structure without buffer area.

However, it is neither advisable to let $\bar{b}$ be a tremendous value to guarantee its accuracy near 0 because a large buffer means a remote distance between the two sections. With such remote distance, when SEV's type changes or LM runs into an incorrect section, its self-adaptation ability degenerates, and point-tracking speed is lowered.

Thus, a buffer mechanism is necessary for SASSB to guarantee its efficiency when locating a target point near 0, as long as $\bar{b}$ is fixedly set in a reasonable range.

4.3.3 Simulation Studies

This section presents a series of experimental results to demonstrate the performance of SASSB in dual environments, in which ASS and SHSSL are added to the comparisons.

Note that in [38], the designed competition between ASS and HSSL is based on state transition time which is unfair to ASS because, in each state transition, HSSL must obtain three new feedbacks from SEV while ASS obtains one only. Therefore, they should be given the same amount of resources during the same period. Thus we use an epoch to represent a time unit during which there is only one feedback, i.e., one epoch means that an LM receives one feedback from SEV.

Based on the above principle, the experimental results among ASS, SHSSL, and SASSB with different $\hat{p}$ values are shown in Figs. 4.25–4.30 where $\bar{b} = 1$.

In Figs. 4.25–4.27, $\hat{p} > 0.5$. It is notable that after adopting the above comparison standard, ASS and SASSB show their excellent search performance. In the aspect of convergence speed, the pair of ASS-related schemes present a strong leading position. It can also be observed that even if its search structure is lengthened by $2b$, SASSB loses a minimal convergence speed as compared with ASS.

In Figs. 4.28–4.30, $\hat{p} < 0.5$. In such deceptive environments, the results show that ASS fails to converge while SASSB performs correctly by maintaining the current speed priority over SHSSL in an informative environment.

In conclusion, all the experimental results show that SASSB inherits the merits of ASS while completely avoiding the latter's failure to handle a deceptive environment. Meanwhile, SASSB is demonstrated to be faster than SHSSL in dealing with dual environments, thereby making it the fastest one in such environments to the authors' best knowledge.

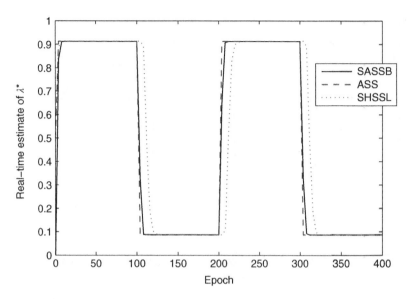

Figure 4.25 Case of $\hat{p} = 0.95$.

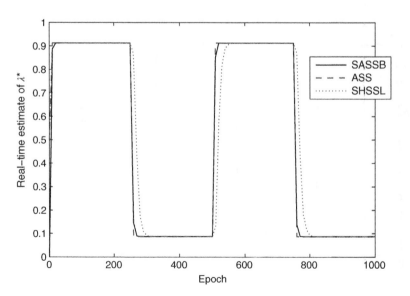

Figure 4.26 Case of $\hat{p} = 0.85$.

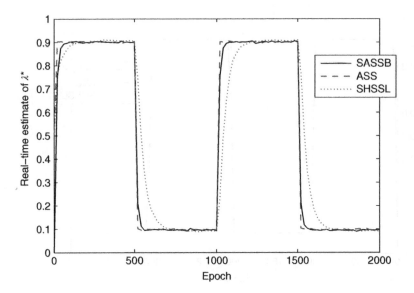

Figure 4.27 Case of $\hat{p} = 0.70$.

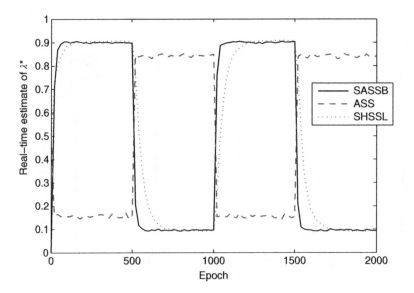

Figure 4.28 Case of $\hat{p} = 0.30$.

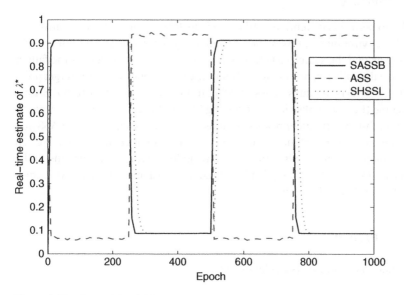

Figure 4.29 Case of $\hat{p} = 0.15$.

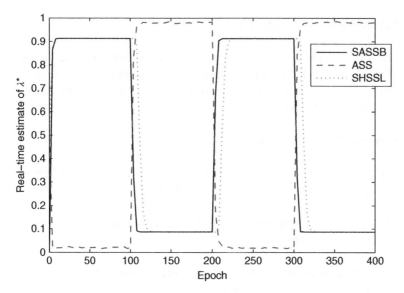

Figure 4.30 Case of $\hat{p} = 0.05$.

4.3.4 Summary

In order to perform conventional Adaptive Step Searches in both informative and deceptive environments, this section combines symmetrization and buffer techniques to extend the search structure of ASS. This combination avoids the failure of ASS in a deceptive environment and guarantees efficiency when the target point is near 0. The existing advantages of ASS are fully retained. As a result, SASSB has become the fastest algorithm for solving SPL problems in dual environments. A series of fair experimental results demonstrate SASSB's superiority to ASS and SHSSL for finding both static and dynamic points in dual environments. We can clearly conclude that SASSB is the best scheme for solving SPL problems to the authors' best knowledge. Its applications to compex systems, especially cyberphysical systems [43–47] should be pursued.

4.4 Exercises

1 What types can events be classified?

2 What affects the efficiency of the whole recognition system?

3 What is the difference between FSSA and VSSA?

4 What are the significant advantages of STPLA?

5 Using STPLA with an omission error of 0.2, calculate the formula for the probability that the learning automaton will successfully determine the existence of a pattern and analyze the necessity of accelerating the convergence process based on the results.

6 Design an AT-STPLA with 5-tuple (A,B,Q,T,G).

7 Calculate the minimum value of $\bar{s}(t)$ when q(t) arrives at state one from a non-one state during the state transition of AT-STPLA.

8 How does the alert probability change when the occurrence probability of an event in AT-STPLA changes?

9 Which phase does the AT-STPLA differentiate from other variants of Spatio-Temporal learning automata?

10 What are the flaws of Hierarchical stochastic searching?

11 Design the search structure of HSSL, $\overline{D} = 3$.

12 Assume the target parameter is set as 0.9123, $\overline{D} = 3$, $\hat{p} = 0.3$, and the initial search node is set as $\overline{S}_{\{0,1\}}$, Please trace the execution process and determine when it converges.

13 What is the difference between ASS and traditional SPL methods?

14 How does the ASS perform Adaptive Step Searches in informative and deceptive environments?

15 Why a too large buffer is not advisable to SASSB?

References

1 W. Duo, M. Zhou, and A. Abusorrah, "A survey of cyber attacks on cyber physical systems: Recent advances and challenges," *IEEE/CAA Journal of Automatica Sinica*, vol. 9, no. 5, pp. 784–800, 2022.

2 B. Huang, C. W. M. Zhou, A. Abusorrah, and Y. Al-Turki, "Formal modeling and discovery of multi-instance business processes: A cloud resource management case study," *IEEE/CAA Journal of Automatica Sinica*, vol. 8, no. 3, pp. 597–605, 2021.

3 C. Liu, "Dual-environmental particle swarm optimizer in noisy and noise-free environments," *IEEE/CAA Journal of Automatica Sinica*, vol. 9, no. 12, pp. 2151–2160, 2022.

4 X. Wang, H. Hu, and M. Zhou, "Discrete event approach to robust control in automated manufacturing systems," *IEEE Transactions on Systems, Man, and Cybernetics: Systems*, vol. 52, no. 1, pp. 123–135, 2022.

5 A. Yazidi, O.-C. Granmo, and B. J. Oommen, "Learning automaton based on-line discovery and tracking of spatio-temporal event patterns," *IEEE Transactions on Cybernetics*, vol. 43, no. 3, pp. 1118–1130, 2013.

6 P. Rashidi and D. J. Cook, "Keeping the resident in the loop: Adapting the smart home to the user," *IEEE Transactions on Systems, Man, and Cybernetics - Part A: Systems and Humans*, vol. 39, no. 5, pp. 949–959, 2009.

7 J. Zhang, Y. Wang, C. Wang, and M. C. Zhou, "Fast variable structure stochastic automaton for discovering and tracking spatiotemporal event patterns," *IEEE Transactions on Cybernetics*, vol. 48, no. 3, pp. 890–903, 2018.

8 W. Jiang, C. L. Zhao, S. H. Li, and L. Chen, "A new learning automata based approach for online tracking of event patterns," *Neurocomputing*, vol. 137, no. 11, pp. 205–211, 2014.

9 A. H. Jamalian, R. Iraji, A. R. Sefidpour, and M. T. Manzuri-Shalmani, "Examining the ε-optimality property of a tunable FSSA," in *6th IEEE International Conference on Cognitive Informatics*, IEEE, 2007, pp. 169–177.

10 M. L. Tsetlin, "On the behavior of finite automata in random media," *Automation and Remote Control*, vol. 22, no. 10, pp. 1345–1354, 1961.

11 M. L. Tsetlin, "Automaton theory and modeling of biological systems," *American Economic Review*, vol. 102, pp. 234–244, 1973.

12 H. Sun, V. D. Florio, N. Gui, and C. Blondia, "Promises and challenges of ambient assisted living systems," in *6th International Conference on Information Technology: New Generations*, 2009, pp. 1201–1207.

13 N. Shiryaev, "Probability," (2nd ed.). New York: Springer-Verlag, 1995.

14 J. Zhang, X. Zhu, Y. Wang, and M. Zhou, "Symmetrical hierarchical stochastic searching on the line in informative and deceptive environments," *IEEE Transactions on Cybernetics*, vol. 47, no. 3, pp. 626–635, 2017.

15 M. Livio, "The golden ratio: The story of phi, the world's most astonishing number," Portland, OR: Broadway Books, 2008.

16 A. Yazidi, O.-C. Granmo, B. J. Oommen, and M. Goodwin, "A novel strategy for solving the stochastic point location problem using a hierarchical searching scheme," *IEEE Transactions on Cybernetics*, vol. 44, no. 11, pp. 2202–2220, 2014.

17 B. J. Oommen, "Stochastic searching on the line and its applications to parameter learning in nonlinear optimization," *IEEE Transactions on Systems, Man, and Cybernetics, Part B (Cybernetics)*, vol. 27, no. 4, pp. 733–739, 1997.

18 B. J. Oommen, G. Raghunath, and B. Kuipers, "Parameter learning from stochastic teachers and stochastic compulsive liars," *IEEE Transactions on Systems, Man, and Cybernetics, Part B (Cybernetics)*, vol. 36, no. 4, pp. 820–834, 2006.

19 B. J. Oommen, S.-W. Kim, M. T. Samuel, and O.-C. Granmo, "A solution to the stochastic point location problem in metalevel nonstationary environments," *IEEE Transactions on Systems, Man, and Cybernetics, Part B (Cybernetics)*, vol. 38, no. 2, pp. 466–476, 2008.

20 S. Lakshmivarahan, "Learning algorithms theory and applications," New York: Springer-Verlag, 1981.

21 K. Najim and A. S. Poznyak, "Learning automata: Theory and applications," New York: Pergamon, 1994.

22 K. S. Narendra and M. A. L. Thathachar, "Learning automata: An introduction," Englewood Cliffs, NJ: Prentice-Hall, 1989.

23 M. S. Obaidat, G. I. Papadimitriou, and A. S. Pomportsis, "Guest editorial learning automata: Theory, paradigms, and applications," *IEEE Transactions on Systems, Man, and Cybernetics, Part B (Cybernetics)*, vol. 32, no. 6, pp. 706–709, 2002.

24 S. Misra, V. Krishna, V. Saritha, and M. S. Obaidat, "Learning automata as a utility for power management in smart grids," *IEEE Communications Magazine*, vol. 51, no. 1, pp. 98–104, 2013.

25 Y. Xu, J. Wang, Q. Wu, A. Anpalagan, and Y. Yao, "Opportunistic spectrum access in unknown dynamic environment: A game-theoretic stochastic learning solution," *IEEE Transactions on Wireless Communications*, vol. 11, no. 4, pp. 1380–1391, 2012.

26 J. A. Torkestani and M. R. Meybodi, "Finding minimum weight connected dominating set in stochastic graph based on learning automata," *Information Sciences*, vol. 200, no. 1, pp. 57–77, 2012.

27 S. Afshar, M. Mosleh, and M. Kheyrandish, "Presenting a new multiclass classifier based on learning automata," *Neurocomputing*, vol. 104, pp. 97–104, 2013.

28 P. Zhou, Y. Chang, and J. A. Copeland, "Reinforcement learning for repeated power control game in cognitive radio networks," *IEEE Journal on Selected Areas in Communications*, vol. 30, no. 1, pp. 54–69, 2012.

29 A. Yazidi, O.-C. Granmo, and B. J. Oommen, "Service selection in stochastic environments: A learning-automaton based solution," *Applied Intelligence*, vol. 36, no. 3, pp. 617–637, 2012.

30 P. Nicopolitidis, G. Papadimitriou, A. Pomportsis, P. Sarigiannidis, and M. Obaidat, "Adaptive wireless networks using learning automata," *IEEE Wireless Communications*, vol. 18, no. 2, pp. 75–81, 2011.

31 O. Tilak, R. Martin, and S. Mukhopadhyay, "Decentralized indirect methods for learning automata games," *IEEE Transactions on Systems, Man, and Cybernetics, Part B (Cybernetics)*, vol. 41, no. 5, pp. 1213–1223, 2011.

32 P. Nicopolitidis, V. Kakali, G. Papadimitriou, and A. Pomportsis, "On performance improvement of wireless push systems via smart antennas," *IEEE Transactions on Communications*, vol. 60, no. 2, pp. 312–316, 2012.

33 T. Ben-Zvi and J. Nickerson, "Decision analysis: Environmental learning automata for sensor placement," *IEEE Sensors Journal*, vol. 11, no. 5, pp. 1206–1207, 2011.

34 B. J. Oommen, O.-C. Granmo, and Z. Liang, "A novel multidimensional scaling technique for mapping word-of-mouth discussions," in *Studies in Computational Intelligence*, vol. 214, pp. 317–322, Berlin, Heidelberg: Springer-Verlag, 2009.

35 B. J. Oommen and D. Calitoiu, "Modeling and simulating a disease outbreak by learning a contagion parameter-based model," in *Spring Simulation Multiconference*, 2008, pp. 547–555.

36 B. J. Oommen and G. Raghunath, "Automata learning and intelligent tertiary searching for stochastic point location," *IEEE Transactions on Systems, Man, and Cybernetics, Part B (Cybernetics)*, vol. 28, no. 6, pp. 947–954, 1998.

37 D. Huang and W. Jiang, "A general CPL-AdS methodology for fixing dynamic parameters in dual environments," *IEEE Transactions on Systems, Man, and Cybernetics, Part B (Cybernetics)*, vol. 42, no. 5, pp. 1489–1500, 2012.

38 T. Tao, H. Ge, G. Cai, and S. Li, "Adaptive step searching for solving stochastic point location problem," in *International Conference on Intelligent Computing Theories*, Lecture Notes in Computer Science, pp. 192–198, Berlin, Heidelberg: Springer-Verlag, 2013.

39 F. P. Kelly, "*Reversibility and stochastic networks*," Wiley Series in Probability and Mathematical Statistics Tracts on Probability and Statistics. New York: Wiley, 1987.

40 O.-C. Granmo and B. J. Oommen, "Solving stochastic nonlinear resource allocation problems using a hierarchy of twofold resource allocation automata," *IEEE Transactions on Computers*, vol. 59, no. 4, pp. 545–560, 2009.

41 A. Yazidi, O.-C. Granmo, B. J. Oommen, and M. Goodwin, "A hierarchical learning scheme for solving the stochastic point location problem," in *Advanced Research in Applied Artificial Intelligence*, pp. 774–783, Berlin, Heidelberg: Springer-Verlag, 2012.

42 W. Jiang, D.-S. Huang, and S. Li, "Random walk-based solution to triple level stochastic point location problem," *IEEE Transactions on Cybernetics*, vol. 46, no. 6, pp. 1438–1451, 2015.

43 W. Duo, M. Zhou, and A. Abusorrah, "A survey of cyber attacks on cyber physical systems: Recent advances and challenges," *IEEE/CAA Journal of Automatica Sinica*, vol. 9, no. 5, pp. 784–800, 2022.

44 Y. Guo, et al., "Mobile cyber physical systems: Current challenges and future networking applications," *IEEE Access*, vol. 6, pp. 12360–12368, 2018.

45 S. Li, M. Zhou, X. Luo, and Z.-H. You, "Distributed winner-take-all in dynamic networks," *IEEE Transactions on Automatic Control*, vol. 62, no. 2, pp. 577–589, 2017.

46 X. Wang, et al., "Dynamic low-power reconfiguration of real-time systems with periodic and probabilistic tasks," *IEEE Transactions on Automation Science and Engineering*, vol. 12, no. 1, pp. 258–271, 2015.

47 X. Hu, et al., "Emotion-aware cognitive system in multi-channel cognitive radio ad hoc networks," *IEEE Communications Magazine*, vol. 56, no. 4, pp. 180–187, 2018

5

Ordinal Optimization

Discrete-event system simulation is a widely used tool for evaluating and analyzing event-driven systems whose real situations rarely satisfy the assumptions of analytical models [1–4]. The efficiency of a simulation procedure is always a significant concern [5], because performances of discrete-event systems under noise (uncertainty) need to be estimated within a finite number of simulation replications. For example, in an archery competition, it is necessary to use as few arrows or competition times as possible to determine the most accurate archer.

Ordinal optimization, introduced by Ho *et al.*, is an efficient technique for discrete-event system simulation and optimization [6]. One of its tenets is the order, rather than the exact values, of alternative options [7]. Ordinal optimization has great advantages on finding better design(s) or worse one(s). Since the ordination can be obtained much easier than exact values, it has been applied to many fields successfully [8–11]. Actually, ordinal optimization can solve the same problem as learning automata do. However, no literature discusses their relationship and differences. Thus, knowing how ordinal optimization works help understand the learning automata better by putting them together and comparing them.

5.1 Optimal Computing-Budget Allocation

To further enhance the efficiency of ordinal optimization, Chen *et al.* [1, 2] have proposed optimal computing-budget allocation (OCBA). In a noisy environment, OCBA selects the best design with a high probability by allocating a suitable number of simulation replications to each design. With its great advantages, OCBA has been widely used in simulation-based situations, such as circuit design problems [12], air traffic management [13], continuous-variable optimization under uncertainty [14] and simulation-based policy improvement [15]. Besides independent sampling case considered in original OCBA, the correlated case is

Learning Automata and Their Applications to Intelligent Systems, First Edition.
JunQi Zhang and MengChu Zhou.

investigated by Fu *et al.* [16]. Later, OCBA for selecting an optimal subset of top-$\hat{m}$ designs (OCBAm) is proposed by Chen *et al.* [17]. OCBAm is able to select the optimal subset containing top $\hat{m}$ designs out of $\hat{k}$ designs ($\hat{k} > \hat{m} > 1$), rather than the top one design. Inspired by OCBA and OCBAm, an OCBA-m allocation procedure is used to allocate the budget for data collection when data envelopment analysis efficiency is predicted [18]. Lee *et al.* [19] have investigated OCBA in constrained optimization and a new OCBA approach is proposed for selecting the best design with stochastic constraint. Another variant of OCBA for stochastic simulation time is developed by Jia [20], which is applicable to the cases when the simulation time is stochastic due to the random system behavior. Considering computing budget allocated to different designs and unequal resources to different systems, Peng *et al.* [21] present a heuristic two-stage sequential allocation algorithm for selecting the best design. All the above variants of OCBA are proposed to select the better ones or best one, out of all designs.

The approximate probability of selecting the correct best design or candidate is asymptotically maximized according to the following equations:

$$\frac{\mathbb{N}_i(t)}{\mathbb{N}_j(t)} = \left(\frac{\hat{\sigma}_i(t)/(\hat{\delta}_{i,\hat{m}}(t))}{\hat{\sigma}_j(t)/(\hat{\delta}_{j,\hat{m}}(t))} \right)^2, \quad \forall i, j \in \mathbb{Z}_r \text{ and } i \neq j \neq \hat{m}, \tag{5.1}$$

$$\mathbb{N}_{\hat{m}}(t) = \hat{\sigma}_{\hat{m}}(t) \sqrt{\sum_{i=1, i \neq \hat{m}}^r \frac{\mathbb{N}_i^2(t)}{(\hat{\sigma}_i(t))^2}}, \tag{5.2}$$

where $\mathbb{N}_i(t)$ refers to the total number of evaluations assigned to action i at allocation time t, $\hat{d}_i(t)$ and $\hat{\sigma}_i(t)$ store its sample mean and variance, respectively, and $\hat{m}$ stands for the best action with the lowest mean, $\hat{\delta}_{i,\hat{m}}(t) = \hat{d}_i(t) - \hat{d}_{\hat{m}}(t)$.

Let $\hat{k} = \sum \mathbb{N}_i(t) / \sum \frac{\hat{\sigma}_i^2(t)}{\hat{\delta}_{i,\hat{m}}^2(t)}, i \in \mathbb{Z}_r$ and $i \neq \hat{m}$. According to (5.1) and (5.2), we have

$$\mathbb{N}_i(t) = \frac{\hat{\sigma}_i^2(t)}{\hat{\delta}_{i,\hat{m}}^2(t)} \hat{k}, \quad i \in \mathbb{Z}_r \text{ and } i \neq \hat{m}. \tag{5.3}$$

$$\mathbb{N}_{\hat{m}}(t) = \hat{\sigma}_{\hat{m}}(t) \sum_{i=1, i \neq \hat{m}}^r \frac{\hat{\sigma}_i(t)}{\hat{\delta}_{i,\hat{m}}^2(t)} \hat{k}. \tag{5.4}$$

Then, we have

$$\mathbb{N}_{\hat{m}}(t) = \frac{\mathbb{N}(t)\hat{\sigma}_{\hat{m}}(t) \sum_{i=1, i \neq \hat{m}}^r \frac{\hat{\sigma}_i(t)}{\hat{\delta}_{i,\hat{m}}^2(t)}}{\hat{\sigma}_{\hat{m}}(t) \sum_{i=1, i \neq \hat{m}}^r \frac{\hat{\sigma}_i(t)}{\hat{\delta}_{i,\hat{m}}^2(t)} + \sum_{i=1, i \neq \hat{m}}^r \frac{\hat{\sigma}_i^2(t)}{\hat{\delta}_{i,\hat{m}}^2(t)}}$$

$$= \mathbb{N}(t) \sum_{i=1, i \neq \hat{m}}^r \frac{\hat{\sigma}_{\hat{m}}(t)}{\hat{\sigma}_{\hat{m}}(t) + \hat{\sigma}_i(t)}, \tag{5.5}$$

where $\mathbb{N}(t) = \sum_{i=1}^r \mathbb{N}_i(t)$ is the total budget.

For $i \neq \hat{m}$, we have

$$\mathbb{N}_i(t) = \frac{\mathbb{N}(t)\frac{\hat{\sigma}_i^2(t)}{\hat{\delta}_{i,\hat{m}}^2(t)}}{\hat{\sigma}_{\hat{m}}(t)\sum_{j=1,j\neq\hat{m}}^r \frac{\hat{\sigma}_j(t)}{\hat{\delta}_{j,\hat{m}}^2(t)} + \sum_{j=1,j\neq\hat{m}}^r \frac{\hat{\sigma}_j^2(t)}{\hat{\delta}_{j,\hat{m}}^2(t)}}. \tag{5.6}$$

Hence, according to (5.5) and (5.6), the total budget can be allocated to each action. The main purpose of OCBA is to allocate limited computational budget among the actions at time t in a way that maximizes the probability of correctly selecting the optimal action. It should be noted that in different applications, "action" can be "design," "system," "option," etc. Next, we introduce two variants of OCBA.

5.2 Optimal Computing-Budget Allocation for Selection of Best and Worst Designs

5.2.1 Background and Motivation

Selecting both the best and worst designs is motivated by the ranking and selection problems in practice, which requires selecting not only the best design but also the worst one simultaneously.

Example 5.1 Many companies identify the best and worst performing employees. Hence, they can reward the best employee and punish or dismiss the worst employee.

Example 5.2 When investing a layout of a municipal sewage treatment plant in a city [22], the best and worst layout plans should be determined in each assessment index, and then the orders of the other layout plans are evaluated according to their distances to the best and worst ones simultaneously. Finally, the best layout plan of a plant is determined based on the comprehensive indexes. The real cases prove that only the best or worst layout plan set as a target in each index is not enough because the best layout plan of a plant in each index may not be same such that the final result cannot be determined based on the comprehensive indexes. Both the best and worst ones being considered can present a reliable and suitable result in layout planning of a municipal sewage treatment plant. Hence, the philosophy of selecting the best and worst designs indicates tremendous potential.

The OCBA introduced in Section 5.1 fails to address the issue of selecting the best and worst designs efficiently. The need to select them both rapidly given a fixed computing budget has arisen from many applications. This section introduces a

OCBA-based approach for selecting both best and worst designs at the same time. It approximately maximizes a lower bound for the probability of correctly selecting the best and worst designs simultaneously.

5.2.2 Approximate Optimal Simulation Budget Allocation

This section requires the following notations:

$\bar{T}$ Total number of simulation replications (the given computing budget);

$\hat{k}$ Total number of designs, satisfying $\hat{k} \in \mathbb{Z}^+ = \{1, 2, \dots\}$ and $\hat{k} \geq 3$;

Θ The set of designs, $\Theta = \{1, 2, \dots, \hat{k}\}$;

$\hat{i}$ Design $\hat{i}$, the best one to be selected in the set of designs;

$\check{i}$ Design $\check{i}$, the worst one to be selected in the set of designs;

$\mathbb{N}_i$ The number of simulation replications allocated to design $i \in \Theta$, satisfying $\mathbb{N}_i \in \mathbb{Z}^+$;

$\hat{J}_{ij}$ The output of the jth simulation replication for design $i \in \Theta$, $j = 1, 2, \dots, \mathbb{N}_i$;

$\bar{J}_i$ The sample mean for design $i \in \Theta$;

$\hat{\sigma}_i^2$ The variance for design $i \in \Theta$. It is unknown in practice beforehand and so is approximated by sample variance;

J_i The mean for design $i \in \Theta$. It is unknown in practice beforehand and so is approximated by sample mean;

$\tilde{J}_i$ The posterior estimate of J_i;

$\tilde{\varphi}(x)$ The standard normal probability density function, $\tilde{\varphi}(x) = \frac{1}{\sqrt{2\pi}} e^{-x^2/2}$;

$\tilde{\Phi}(x)$ The standard normal cumulative distribution function, $\tilde{\Phi}(x) = \int_{-\infty}^{x} \tilde{\varphi}(t) dt$.

Considering a discrete-event system with $\hat{k} \in \mathbb{Z}^+$ designs, the goal is to correctly select the best and worst designs simultaneously. J_i, the true performance of design i, can be accurately evaluated only by an infinite number of simulation replications. That is

$$J_i = \lim_{\mathbb{N}_i \to +\infty} \frac{1}{\mathbb{N}_i} \sum_{j=1}^{\mathbb{N}_i} \hat{J}_{ij}.$$

$\hat{J}_{ij}$, the output of the jth simulation replication for design i, can be calculated as:

$$\hat{J}_{ij} = J_i + \check{i}_{ij},$$

where $\check{i}_{ij}$ follows an i.i.d. (independent and identically distributed) normal distribution $N(0, \hat{\sigma}_i^2)$ for all j to the same i. When $\mathbb{N}_i$ is large enough, the sample mean can closely approximate its true value of design i. The estimate of J_i can be calculated as:

$$\bar{J}_i = \frac{1}{N_i} \sum_{j=1}^{N_i} \tilde{J}_{ij}.$$

Taking the Bayesian viewpoint, the true performance J_i has the posterior estimate $\tilde{J}_i$. In [23], DeGroot shows that

$$\tilde{J}_i \sim N\left(\bar{J}_i, \frac{\hat{\sigma}_i^2}{N_i}\right).$$

Sorting the designs from the small to the large according to $\bar{J}_i$, we let $\hat{i}$ and $\check{i}$ denote the best and worst designs, respectively, i.e., $\bar{J}_{\hat{i}}$ has the smallest value and $\bar{J}_{\check{i}}$ has the largest one. Similar to the simulation work in [1, 2], the *Probability of Correct Selection* denoted by $P\{CS\}$, is defined as the probability to correctly select the best and worst designs simultaneously. Mathematically,

$$P\{CS\} = P\left\{\tilde{J}_{\hat{i}} \leq \tilde{J}_i \leq \tilde{J}_{\check{i}}, \, \forall i \neq \hat{i}, \check{i}\right\}. \tag{5.7}$$

The number of total budgets $\bar{T}$ is given a finite number, satisfying

$$\bar{T} = \sum_{i \in \Theta} N_i, \, N_i \in \mathbb{Z}^+. \tag{5.8}$$

How to maximize $P\{CS\}$ for a given $\bar{T}$ is of interest. In other words, determining suitable N_i to maximize $P\{CS\}$ subject to $\bar{T}$ can be expressed mathematically as follows:

$$\max P\{CS\},$$
$$\text{s.t.} \sum_{i \in \Theta} N_i = \bar{T} \quad \text{and} \quad N_i \in \mathbb{Z}^+. \tag{5.9}$$

$P\{CS\}$ can be estimated by using Monte Carlo simulation, which is a time-consuming procedure due to their massive search space and evolution in time under the influence of random occurrences [24]. Thus, a lower bound can be provided for $P\{CS\}$, i.e.,

$$P\{CS\} = P\left\{\tilde{J}_{\hat{i}} \leq \tilde{J}_i \leq \tilde{J}_{\check{i}}, \, \forall i \neq \hat{i}, \check{i}\right\}$$

$$= P\left\{\bigcap_{i \neq \hat{i}, \check{i}} \left(\tilde{J}_{\hat{i}} \leq \tilde{J}_i \leq \tilde{J}_{\check{i}}\right)\right\}$$

$$\geq 1 - \sum_{i \neq \hat{i}, \check{i}} P\left\{\tilde{J}_{\hat{i}} > \tilde{J}_i\right\} - \sum_{i \neq \hat{i}, \check{i}} P\left\{\tilde{J}_i > \tilde{J}_{\check{i}}\right\}$$

$$= APCS. \tag{5.10}$$

The lower bound of $P\{CS\}$ is referred to as the *Approximate Probability of Correct Selection (APCS)*, similar to those in [1, 2]. *APCS*, as a relatively efficient way of estimating $P\{CS\}$, can be computed much easier than $P\{CS\}$. Moreover, numerical tests show that its use can still lead to efficient procedures [1, 2, 24, 25]. Hence, it is highly preferred to calculate *APCS* instead of $P\{CS\}$. For simplification as done similarly in [1, 2], we temporarily assume that N_i is continuous. Thus, the maximization problem of (5.9) can be switched to the following problem:

$$\max_{N_1,\ldots,N_k} APCS,$$

$$\text{s.t.} \sum_{i\in\Theta} N_i = \bar{T}. \tag{5.11}$$

Let $N_i = \hat{\alpha}_i \bar{T}$. $\hat{\alpha}_i$ can be regarded as a proportion of the total number of simulation replications (the given computing budget) for the ith design. Thus, $\hat{\alpha}_i \in [0, 1]$ and $\sum_{i\in\Theta}\hat{\alpha}_i = 1$.

Our objective is to solve the maximization problem of (5.11) by introducing a Lagrange multiplier $\bar{\lambda}$. From (5.10), we have

$$\bar{F} = 1 - \sum_{i\neq\hat{i},\check{i}}^{\hat{k}} P\left\{\tilde{J}_{\hat{i}} > \tilde{J}_i\right\}$$

$$- \sum_{i\neq\hat{i},\check{i}}^{\hat{k}} P\left\{\tilde{J}_i > \tilde{J}_{\check{i}}\right\} + \bar{\lambda}\left(\sum_{i=1}^{\hat{k}} N_i - \bar{T}\right). \tag{5.12}$$

We consider that $\tilde{J}_i$ is mutually independent. Thus, the distribution of $\tilde{J}_{\hat{i}} - \tilde{J}_i$ and $\tilde{J}_i - \tilde{J}_{\check{i}}$ is as follows:

$$\tilde{J}_{\hat{i}} - \tilde{J}_i \sim N(\hat{\delta}_{\hat{i}i}, \hat{\sigma}_{\hat{i}i}^2), \quad \forall i \neq \hat{i}, \tag{5.13}$$

$$\tilde{J}_i - \tilde{J}_{\check{i}} \sim N(\hat{\delta}_{i\check{i}}, \hat{\sigma}_{i\check{i}}^2), \quad \forall i \neq \check{i}, \tag{5.14}$$

where for notational simplification, we introduce two new notations, i.e., $\hat{\delta}_{ij} := \tilde{J}_i - \tilde{J}_j$ and $\hat{\sigma}_{ij} := \sqrt{\frac{\hat{\sigma}_i^2}{N_i} + \frac{\hat{\sigma}_j^2}{N_j}}$.

Then from (5.13) we can obtain

$$P\left\{\tilde{J}_{\hat{i}} > \tilde{J}_i\right\} = P\left\{\tilde{J}_{\hat{i}} - \tilde{J}_i > 0\right\}$$

$$= \int_0^{+\infty} \frac{1}{\sqrt{2\pi}\hat{\sigma}_{\hat{i}i}} e^{-\frac{(x-\hat{\delta}_{\hat{i}i})^2}{2\hat{\sigma}_{\hat{i}i}^2}} dx$$

$$= \int_{-\frac{\hat{\delta}_{\hat{i}i}}{\hat{\sigma}_{\hat{i}i}}}^{+\infty} \tilde{\varphi}(x) dx = \tilde{\Phi}\left(\frac{\hat{\delta}_{\hat{i}i}}{\hat{\sigma}_{\hat{i}i}}\right). \tag{5.15}$$

Similarly, from (5.14) we have

$$P\left\{\tilde{J}_i > \tilde{J}_{\check{i}}\right\} = \tilde{\Phi}\left(\frac{\hat{\delta}_{i\check{i}}}{\hat{\sigma}_{i\check{i}}}\right). \tag{5.16}$$

According to (5.15) and (5.16), we rewrite (5.12) as

$$\bar{F} = 1 - \sum_{i\neq\hat{i},\check{i}}\tilde{\Phi}\left(\frac{\hat{\delta}_{i\hat{i}}}{\hat{\sigma}_{i\hat{i}}}\right)$$

$$- \sum_{i\neq\hat{i},\check{i}}\tilde{\Phi}\left(\frac{\hat{\delta}_{i\check{i}}}{\hat{\sigma}_{i\check{i}}}\right) + \bar{\lambda}\left(\sum_{i=1}^{k}\mathbb{N}_i - \bar{T}\right). \tag{5.17}$$

Furthermore, the Karush–Kuhn–Tucker (KKT) [26] conditions of this problem to maximize (5.17) can be stated as follows:

$$\frac{\partial\bar{F}}{\partial\mathbb{N}_i} = -\frac{1}{2\sqrt{2\pi}}e^{-\frac{\delta_{i\hat{i}}^2}{2\hat{\sigma}_{i\hat{i}}^2}}\cdot\frac{\hat{\delta}_{i\hat{i}}\hat{\sigma}_i^2}{\hat{\sigma}_{i\hat{i}}^3\mathbb{N}_i^2}$$

$$-\frac{1}{2\sqrt{2\pi}}e^{-\frac{\delta_{i\check{i}}^2}{2\hat{\sigma}_{i\check{i}}^2}}\cdot\frac{\hat{\delta}_{i\check{i}}\hat{\sigma}_i^2}{\hat{\sigma}_{i\check{i}}^3\mathbb{N}_i^2} + \bar{\lambda} = 0, \quad \forall i \notin \{\hat{i},\check{i}\}. \tag{5.18}$$

$$\frac{\partial\bar{F}}{\partial\mathbb{N}_{\hat{i}}} = -\frac{1}{2\sqrt{2\pi}}\sum_{i\neq\hat{i},\check{i}}e^{-\frac{\delta_{i\hat{i}}^2}{2\hat{\sigma}_{i\hat{i}}^2}}\cdot\frac{\hat{\delta}_{i\hat{i}}\hat{\sigma}_{\hat{i}}^2}{\hat{\sigma}_{i\hat{i}}^3\mathbb{N}_{\hat{i}}^2} + \bar{\lambda} = 0, \tag{5.19}$$

$$\frac{\partial\bar{F}}{\partial\mathbb{N}_{\check{i}}} = -\frac{1}{2\sqrt{2\pi}}\sum_{i\neq\hat{i},\check{i}}e^{-\frac{\delta_{i\check{i}}^2}{2\hat{\sigma}_{i\check{i}}^2}}\cdot\frac{\hat{\delta}_{i\check{i}}\hat{\sigma}_{\check{i}}^2}{\hat{\sigma}_{i\check{i}}^3\mathbb{N}_{\check{i}}^2} + \bar{\lambda} = 0. \tag{5.20}$$

From (5.18)–(5.20), we have

$$-\frac{1}{2\sqrt{2\pi}}e^{-\frac{\delta_{i\hat{i}}^2}{2\hat{\sigma}_{i\hat{i}}^2}}\cdot\frac{\hat{\delta}_{i\hat{i}}}{\hat{\sigma}_{i\hat{i}}^3} - \frac{1}{2\sqrt{2\pi}}e^{-\frac{\delta_{i\check{i}}^2}{2\hat{\sigma}_{i\check{i}}^2}}\cdot\frac{\hat{\delta}_{i\check{i}}}{\hat{\sigma}_{i\check{i}}^3} = -\frac{\bar{\lambda}\mathbb{N}_i^2}{\hat{\sigma}_i^2}, \tag{5.21}$$

$$-\frac{1}{2\sqrt{2\pi}}\sum_{i\neq\hat{i},\check{i}}e^{-\frac{\delta_{i\hat{i}}^2}{2\hat{\sigma}_{i\hat{i}}^2}}\cdot\frac{\hat{\delta}_{i\hat{i}}}{\hat{\sigma}_{i\hat{i}}^3} = -\frac{\bar{\lambda}\mathbb{N}_{\hat{i}}^2}{\hat{\sigma}_{\hat{i}}^2}, \tag{5.22}$$

$$-\frac{1}{2\sqrt{2\pi}}\sum_{i\neq\hat{i},\check{i}}e^{-\frac{\delta_{i\check{i}}^2}{2\hat{\sigma}_{i\check{i}}^2}}\cdot\frac{\hat{\delta}_{i\check{i}}}{\hat{\sigma}_{i\check{i}}^3} = -\frac{\bar{\lambda}\mathbb{N}_{\check{i}}^2}{\hat{\sigma}_{\check{i}}^2}, \tag{5.23}$$

Then, three relationships are next inspected in terms of the number of simulation replications: the relationship among $\mathbb{N}_{\hat{i}}$, $\mathbb{N}_{\check{i}}$, and $\mathbb{N}_i$, $i \neq \hat{i},\check{i}$, the relationship among $\mathbb{N}_i$, $i \neq \hat{i},\check{i}$, and the relationship between $\mathbb{N}_{\hat{i}}$ and $\mathbb{N}_{\check{i}}$. We inspect them one by one.

From (5.21)–(5.23),

$$\sum_{i\neq \hat{i},\hat{i}} \frac{\bar{\lambda}\mathbb{N}_i^2}{\hat{\sigma}_i^2} = \frac{\bar{\lambda}\mathbb{N}_i^2}{\hat{\sigma}_i^2} + \frac{\bar{\lambda}\mathbb{N}_{\tilde{i}}^2}{\hat{\sigma}_{\tilde{i}}^2}. \tag{5.24}$$

Thus, the first relationship can be expressed from (5.24) as:

$$\frac{\mathbb{N}_i^2}{\hat{\sigma}_i^2} + \frac{\mathbb{N}_{\tilde{i}}^2}{\hat{\sigma}_{\tilde{i}}^2} = \sum_{i\neq \hat{i},\tilde{i}} \frac{\mathbb{N}_i^2}{\hat{\sigma}_i^2}. \tag{5.25}$$

Before investigating the relationship among $\mathbb{N}_i$, the following lemma can be proved.

Lemma 5.1 *If* $\sum_{i=1}^{\hat{k}} \dot{A}_i e^{\dot{a}_i x}$ *is in the same order with* $\sum_{j=1}^{\hat{k}} \dot{B}_j e^{\dot{b}_j x}$ *as x increases*

$$\lim_{x\to+\infty} \frac{\sum_{i=1}^{\hat{k}} \dot{A}_i e^{\dot{a}_i x}}{\sum_{j=1}^{\hat{k}} \dot{B}_j e^{\dot{b}_j x}} = r(r > 0) \tag{5.26}$$

then

$$\max_i \dot{a}_i = \max_j \dot{b}_j.$$

Proof: When $x \to +\infty$, the relationship between $\sum_{i=1}^{\hat{k}} \dot{A}_i e^{\dot{a}_i x}$ and $\sum_{j=1}^{\hat{k}} \dot{B}_j e^{\dot{b}_j x}$ is

$$\lim_{x\to+\infty} \frac{\sum_{i=1}^{\hat{k}} \dot{A}_i \cdot e^{\dot{a}_i x}}{\sum_{j=1}^{\hat{k}} \dot{B}_j \cdot e^{\dot{b}_j x}} = \lim_{x\to+\infty} \frac{\dot{A}_{i^*} \cdot e^{\dot{a}_{i^*} x} + \sum_{i=1, i\neq i^*}^{\hat{k}} \dot{A}_i \cdot e^{\dot{a}_i x}}{\dot{B}_{j^*} \cdot e^{\dot{b}_{j^*} x} + \sum_{j=1, j\neq j^*}^{\hat{k}} \dot{B}_j \cdot e^{\dot{b}_j x}},$$

where $i^* \in \arg\max_i \dot{a}_i$ and $j^* \in \arg\max_j \dot{b}_j$. Meanwhile, $\sum_{i=1, i\neq i^*}^{\hat{k}} \dot{A}_i \cdot e^{\dot{a}_i x} = o(\dot{A}_{i^*} \cdot e^{\dot{a}_{i^*} x})$ and $\sum_{j=1, j\neq j^*}^{\hat{k}} \dot{B}_j \cdot e^{\dot{b}_j x} = o(\dot{B}_{j^*} \cdot e^{\dot{b}_{j^*} x})$. Thus we have the first conclusion:

$$\lim_{x\to+\infty} \frac{\sum_{i=1}^{\hat{k}} \dot{A}_i \cdot e^{\dot{a}_i x}}{\sum_{j=1}^{\hat{k}} \dot{B}_j \cdot e^{\dot{b}_j x}} = \lim_{x\to+\infty} \frac{\dot{A}_{i^*} \cdot e^{\dot{a}_{i^*} x}}{\dot{B}_{j^*} \cdot e^{\dot{b}_{j^*} x}}.$$

Based on the assumption from (5.26):

$$\lim_{x\to+\infty} \frac{\sum_{i=1}^{\hat{k}} \dot{A}_i e^{\dot{a}_i x}}{\sum_{j=1}^{\hat{k}} \dot{B}_j e^{\dot{b}_j x}} = r(r > 0),$$

we have

$$\lim_{x\to+\infty} \frac{\dot{A}_{i^*} \cdot e^{\dot{a}_{i^*} x}}{\dot{B}_{j^*} \cdot e^{\dot{b}_{j^*} x}} = r.$$

We assume that $\dot{a}_{i^*} \neq \dot{b}_{j^*}$, i.e., $\dot{a}_{i^*} > \dot{b}_{j^*}$ or $\dot{a}_{i^*} < \dot{b}_{j^*}$. In these two cases,

$$\lim_{x \to +\infty} \frac{\sum_{i=1}^{k} \dot{A}_i e^{\dot{a}_i x}}{\sum_{j=1}^{k} \dot{B}_j e^{\dot{b}_j x}} \to \infty \; or \; 0,$$

which violates the assumption of Lemma 5.1. Thus, we can further draw the second conclusion:

$$\lim_{x \to +\infty} \frac{\dot{A}_{i^*} \cdot e^{\dot{a}_{i^*} x}}{\dot{B}_{j^*} \cdot e^{\dot{b}_{j^*} x}} = r \Rightarrow \dot{a}_{i^*} = \dot{b}_{j^*} \Rightarrow \max_i \dot{a}_i = \max_j \dot{b}_j.$$

where $i^* \in \arg\max_i \dot{a}_i$ and $j^* \in \arg\max_j \dot{b}_j$. ∎

From (5.25), we have $\frac{N_{\hat{i}}^2}{\hat{\sigma}_{\hat{i}}^2} + \frac{N_{\check{i}}^2}{\hat{\sigma}_{\check{i}}^2} \gg \frac{N_i^2}{\hat{\sigma}_i^2}$, $i \neq \hat{i}, \check{i}$. When $\hat{\sigma}_1 = \hat{\sigma}_2 = \cdots = \hat{\sigma}_{\hat{k}}$,

$$N_{\hat{i}} + N_{\check{i}} \geq \sqrt{N_{\hat{i}}^2 + N_{\check{i}}^2} = \sqrt{\sum_{i \neq \hat{i}, \check{i}}^{\hat{k}} N_i^2}.$$

As $\hat{k}$ increases, the sum of the number of simulation replications allocated to the best and worst designs increases relatively. Therefore, the assumptions $N_{\hat{i}} \gg N_i$ and $N_{\check{i}} \gg N_i$ are reasonable to further simplify the approximation.

Thus, $\hat{\sigma}_{\hat{i}i}^2$ and $\hat{\sigma}_{\check{i}i}^2$ can be approximated as:

$$\hat{\sigma}_{\hat{i}i}^2 = \frac{\hat{\sigma}_{\hat{i}}^2}{N_{\hat{i}}} + \frac{\hat{\sigma}_i^2}{N_i} \approx \frac{\hat{\sigma}_i^2}{N_i}, \tag{5.27}$$

$$\hat{\sigma}_{\check{i}i}^2 = \frac{\hat{\sigma}_{\check{i}}^2}{N_{\check{i}}} + \frac{\hat{\sigma}_i^2}{N_i} \approx \frac{\hat{\sigma}_i^2}{N_i}. \tag{5.28}$$

Based on (5.27) and (5.28), the relationship between designs i and j can be obtained from (5.21),

$$e^{-\frac{\hat{\delta}_{\hat{i}i}^2}{2\hat{\sigma}_{\hat{i}i}^2}} \cdot \frac{\hat{\delta}_{\hat{i}i}\hat{\sigma}_i^2}{\hat{\sigma}_{\hat{i}i}^3 N_i^2} + e^{-\frac{\hat{\delta}_{\check{i}i}^2}{2\hat{\sigma}_{\check{i}i}^2}} \cdot \frac{\hat{\delta}_{\check{i}i}\hat{\sigma}_i^2}{\hat{\sigma}_{\check{i}i}^3 N_i^2}$$

$$= e^{-\frac{\hat{\delta}_{\hat{i}j}^2}{2\hat{\sigma}_{\hat{i}j}^2}} \cdot \frac{\hat{\delta}_{\hat{i}j}\hat{\sigma}_j^2}{\hat{\sigma}_{\hat{i}j}^3 N_j^2} + e^{-\frac{\hat{\delta}_{\check{j}i}^2}{2\hat{\sigma}_{\check{j}i}^2}} \cdot \frac{\hat{\delta}_{\check{j}i}\hat{\sigma}_j^2}{\hat{\sigma}_{\check{j}i}^3 N_j^2}.$$

Now, we consider $\bar{T} \to +\infty$, a widely accepted assumption in the field of ordinal optimization although it is impossible to reach such conditions in real life. The above expression can be simplified as

$$e^{-\frac{\hat{\delta}_{\hat{i}i}^2 \dot{a}_i \bar{T}}{2\hat{\sigma}_i^2}} \cdot \frac{\hat{\delta}_{\hat{i}i}}{\hat{\sigma}_i \dot{a}_i^{\frac{1}{2}}} + e^{-\frac{\hat{\delta}_{\check{i}i}^2 \dot{a}_i \bar{T}}{2\hat{\sigma}_i^2}} \cdot \frac{\hat{\delta}_{\check{i}i}}{\hat{\sigma}_i \dot{a}_i^{\frac{1}{2}}}$$

$$= e^{-\frac{\hat{\delta}_{ij}^2 \hat{a}_j \hat{T}}{2\hat{\sigma}_j^2}} \cdot \frac{\hat{\delta}_{ij}}{\hat{\sigma}_j \hat{\alpha}_j^{\frac{1}{2}}} + e^{-\frac{\hat{\delta}_{ji}^2 \hat{a}_j \hat{T}}{2\hat{\sigma}_j^2}} \cdot \frac{\hat{\delta}_{ji}}{\hat{\sigma}_j \hat{\alpha}_j^{\frac{1}{2}}}.$$

It leads to the following relationship as proved in Lemma 5.1.

$$\max\left(-\frac{\hat{\delta}_{ii}^2 \hat{\alpha}_i}{2\hat{\sigma}_i^2}, -\frac{\hat{\delta}_{ii}^2 \hat{\alpha}_i}{2\hat{\sigma}_i^2}\right) = \max\left(-\frac{\hat{\delta}_{ij}^2 \hat{\alpha}_j}{2\hat{\sigma}_j^2}, -\frac{\hat{\delta}_{ji}^2 \hat{\alpha}_j}{2\hat{\sigma}_j^2}\right)$$

$$\Rightarrow \frac{\hat{\alpha}_i}{\hat{\sigma}_i^2} \min\left(\hat{\delta}_{ii}^2, \hat{\delta}_{ii}^2\right) = \frac{\hat{\alpha}_j}{\hat{\sigma}_j^2} \min\left(\hat{\delta}_{ij}^2, \hat{\delta}_{ji}^2\right). \tag{5.29}$$

From (5.29), the conclusion about the second relationship can be obtained as

$$\frac{N_i}{N_j} = \frac{\hat{\alpha}_i}{\hat{\alpha}_j} = \frac{\hat{\sigma}_i^2 / \min\left(\hat{\delta}_{ii}^2, \hat{\delta}_{ii}^2\right)}{\hat{\sigma}_j^2 / \min\left(\hat{\delta}_{ij}^2, \hat{\delta}_{ji}^2\right)}. \tag{5.30}$$

From (5.19) and (5.20), we obtain

$$\frac{N_i^2 / \hat{\sigma}_i^2}{N_i^2 / \hat{\sigma}_i^2} = \frac{\sum_{i \neq \hat{i}, \hat{i}} e^{-\frac{\hat{\delta}_{ii}^2}{2\hat{\sigma}_{ii}^2}} \cdot \frac{\hat{\delta}_{ii}}{\hat{\sigma}_{ii}^3}}{\sum_{j \neq \hat{i}, \hat{i}} e^{-\frac{\hat{\delta}_{ji}^2}{2\hat{\sigma}_{ji}^2}} \cdot \frac{\hat{\delta}_{ji}}{\hat{\sigma}_{ji}^3}}. \tag{5.31}$$

By the approximation $\hat{\sigma}_{ii}^2 \approx \hat{\sigma}_i^2 / N_i \approx \hat{\sigma}_{ii}^2$ from (5.27) and (5.28), (5.31) can be further simplified into:

$$\frac{N_i^2 / \hat{\sigma}_i^2}{N_i^2 / \hat{\sigma}_i^2} = \frac{\sum_{i \neq \hat{i}, \hat{i}} e^{-\frac{\hat{\delta}_{ii}^2 \hat{a}_i \hat{T}}{2\hat{\sigma}_i^2}} \cdot \hat{\delta}_{ii} \left(\frac{\hat{\alpha}_i}{\hat{\sigma}_i^2}\right)^{\frac{3}{2}}}{\sum_{j \neq \hat{i}, \hat{i}} e^{-\frac{\hat{\delta}_{ji}^2 \hat{a}_j \hat{T}}{2\hat{\sigma}_j^2}} \cdot \hat{\delta}_{ji} \left(\frac{\hat{\alpha}_j}{\hat{\sigma}_j^2}\right)^{\frac{3}{2}}}. \tag{5.32}$$

In the meantime, $\hat{\alpha}_i / \hat{\sigma}_i^2 \propto 1 / \min\left(\hat{\delta}_{ii}^2, \hat{\delta}_{ii}^2\right)$ which means $\hat{\alpha}_i / \hat{\sigma}_i^2$ is proportional to $1 / \min\left(\hat{\delta}_{ii}^2, \hat{\delta}_{ii}^2\right)$, is suggested from the conclusion of (5.30). Therefore, $\hat{\alpha}_i / \hat{\sigma}_i^2$ can be parameterized by $1 / \min\left(\hat{\delta}_{ii}^2, \hat{\delta}_{ii}^2\right)$ and (5.32) can be rewritten as follows:

$$\frac{N_i^2 / \hat{\sigma}_i^2}{N_i^2 / \hat{\sigma}_i^2} = \frac{\sum_{i \neq \hat{i}, \hat{i}} e^{-\frac{\hat{\delta}_{ii}^2 \hat{a}_i \hat{T}}{2\hat{\sigma}_i^2}} \cdot \frac{\hat{\delta}_{ii}}{\min\left(\hat{\delta}_{ii}^3, \hat{\delta}_{ii}^3\right)}}{\sum_{j \neq \hat{i}, \hat{i}} e^{-\frac{\hat{\delta}_{ji}^2 \hat{a}_j \hat{T}}{2\hat{\sigma}_j^2}} \cdot \frac{\hat{\delta}_{ji}}{\min\left(\hat{\delta}_{ij}^3, \hat{\delta}_{ji}^3\right)}}. \tag{5.33}$$

Here, $\dot{\beta}$ and $\dot{\omega}$ are two new notations, defined as

$$\dot{\beta} = \min_{i \neq \hat{i}, \breve{i}} \frac{\hat{\delta}_{\hat{i}\hat{i}}^2 \hat{\alpha}_i}{\hat{\sigma}_i^2} \quad \text{and} \quad \dot{\omega} = \min_{j \neq \hat{i}, \breve{i}} \frac{\hat{\delta}_{\breve{j}\breve{i}}^2 \hat{\alpha}_j}{\hat{\sigma}_j^2}.$$

Two more new notations, $\dot{\theta}_{\dot{\beta}}$ and $\dot{\theta}_{\dot{\omega}}$, are defined as

$$\dot{\theta}_{\dot{\beta}} = \sum_{i \in \left\{ i \mid \frac{\hat{\delta}_{\hat{i}\hat{i}}^2 \hat{\alpha}_i}{\hat{\sigma}_i^2} = \dot{\beta} \right\}} \frac{\hat{\delta}_{\hat{i}\hat{i}}}{\min \left(\hat{\delta}_{\hat{i}\hat{i}}^3, \hat{\delta}_{\hat{i}\hat{i}}^3 \right)},$$

$$\dot{\theta}_{\dot{\omega}} = \sum_{j \in \left\{ j \mid \frac{\hat{\delta}_{\breve{j}\breve{i}}^2 \hat{\alpha}_j}{\hat{\sigma}_j^2} = \dot{\omega} \right\}} \frac{\hat{\delta}_{\breve{j}\breve{i}}}{\min \left(\hat{\delta}_{\hat{i}\hat{j}}^3, \hat{\delta}_{\breve{j}\breve{i}}^3 \right)}.$$

Thus, according to the first conclusion of Lemma 5.1, it can be obtained from (5.31) that: when $\bar{T} \to +\infty$,

$$\frac{\mathbb{N}_{\hat{i}}^2 / \hat{\sigma}_{\hat{i}}^2}{\mathbb{N}_{\breve{i}}^2 / \hat{\sigma}_{\breve{i}}^2} = \frac{\sum_{i \neq \hat{i}, \breve{i}} e^{-\frac{\hat{\delta}_{\hat{i}\hat{i}}^2 \hat{\alpha}_i \bar{T}}{2\hat{\sigma}_i^2}} \cdot \frac{\hat{\delta}_{\hat{i}\hat{i}}}{\min \left(\hat{\delta}_{\hat{i}\hat{i}}^3, \hat{\delta}_{\hat{i}\hat{i}}^3 \right)}}{\sum_{j \neq \hat{i}, \breve{i}} e^{-\frac{\hat{\delta}_{\breve{j}\breve{i}}^2 \hat{\alpha}_j \bar{T}}{2\hat{\sigma}_j^2}} \cdot \frac{\hat{\delta}_{\breve{j}\breve{i}}}{\min \left(\hat{\delta}_{\hat{i}\hat{j}}^3, \hat{\delta}_{\breve{j}\breve{i}}^3 \right)}} = \frac{\dot{\theta}_{\dot{\beta}} e^{-\frac{\dot{\beta}}{2}}}{\dot{\theta}_{\dot{\omega}} e^{-\frac{\dot{\omega}}{2}}}. \tag{5.34}$$

Equation (5.34) conveys that the number of replications allocated to $\hat{i}$ or $\breve{i}$ is related to all $i, i \neq \hat{i}, \breve{i}$. It is approximately satisfied that $\forall i, j \neq \hat{i}, \breve{i}$,

$$\min \left(\frac{\hat{\delta}_{\hat{i}\hat{i}}^2 \hat{\alpha}_i}{\hat{\sigma}_i^2}, \frac{\hat{\delta}_{\breve{i}\breve{i}}^2 \hat{\alpha}_i}{\hat{\sigma}_i^2} \right) = \min \left(\frac{\hat{\delta}_{\hat{i}\hat{j}}^2 \hat{\alpha}_j}{\hat{\sigma}_j^2}, \frac{\hat{\delta}_{\breve{j}\breve{i}}^2 \hat{\alpha}_j}{\hat{\sigma}_j^2} \right). \tag{5.35}$$

Hence, we need to discuss the following three cases according to the smaller one of $\hat{\delta}_{\hat{i}\hat{i}}^2 \hat{\alpha}_i / \hat{\sigma}_i^2$ and $\hat{\delta}_{\breve{i}\breve{i}}^2 \hat{\alpha}_i / \hat{\sigma}_i^2$ for all $i \neq \hat{i}, \breve{i}$

Case 1: If

$$\exists i, \min \left(\frac{\hat{\delta}_{\hat{i}\hat{i}}^2 \hat{\alpha}_i}{\hat{\sigma}_i^2}, \frac{\hat{\delta}_{\breve{i}\breve{i}}^2 \hat{\alpha}_i}{\hat{\sigma}_i^2} \right) = \frac{\hat{\delta}_{\hat{i}\hat{i}}^2 \hat{\alpha}_i}{\hat{\sigma}_i^2}, \tag{5.36}$$

and $\exists j, \min \left(\dfrac{\hat{\delta}_{\hat{i}\hat{j}}^2 \hat{\alpha}_j}{\hat{\sigma}_j^2}, \dfrac{\hat{\delta}_{\breve{j}\breve{i}}^2 \hat{\alpha}_j}{\hat{\sigma}_j^2} \right) = \dfrac{\hat{\delta}_{\breve{j}\breve{i}}^2 \hat{\alpha}_j}{\hat{\sigma}_j^2},$ \hfill (5.37)

then relationship between $\dot{\beta}$ and $\dot{\omega}$ can be obtained

$$\dot{\beta} = \min_{i \neq \hat{i}, \breve{i}} \frac{\hat{\delta}_{\hat{i}\hat{i}}^2 \hat{\alpha}_i}{\hat{\sigma}_i^2} = \min_{j \neq \hat{i}, \breve{i}} \frac{\hat{\delta}_{\breve{j}\breve{i}}^2 \hat{\alpha}_j}{\hat{\sigma}_j^2} = \dot{\omega}.$$

Hence from (5.34), when $\bar{T} \to +\infty$,

$$
\frac{\mathbb{N}_i^2/\hat{\sigma}_i^2}{\mathbb{N}_{\check{\imath}}^2/\hat{\sigma}_{\check{\imath}}^2} = \frac{\dot{\theta}_\beta e^{-\frac{\beta}{2}}}{\dot{\theta}_{\check{\omega}} e^{-\frac{\check{\omega}}{2}}} = \frac{\dot{\theta}_\beta}{\dot{\theta}_{\check{\omega}}}
$$

$$
= \frac{\sum_{i \in \left\{ i \mid \frac{\hat{\delta}_{\hat{\imath}i}^2 \hat{\alpha}_i}{\hat{\sigma}_i^2} = \beta \right\}} \hat{\delta}_{\hat{\imath}i} \bigg/ \min \left(\hat{\delta}_{\hat{\imath}i}^3, \hat{\delta}_{i\hat{\imath}}^3 \right)}{\sum_{j \in \left\{ j \mid \frac{\hat{\delta}_{\check{\imath}j}^2 \hat{\alpha}_j}{\hat{\sigma}_j^2} = \check{\omega} \right\}} \hat{\delta}_{\check{\imath}j} \bigg/ \min \left(\hat{\delta}_{\check{\imath}j}^3, \hat{\delta}_{j\check{\imath}}^3 \right)}. \tag{5.38}
$$

Meanwhile, according to notation

$$
\dot{\beta} = \min_{i \neq \hat{\imath}, \check{\imath}} \frac{\hat{\delta}_{\hat{\imath}i}^2 \hat{\alpha}_i}{\hat{\sigma}_i^2},
$$

it is observed from (5.35) that $\exists i$, when $\hat{\delta}_{\hat{\imath}i}^2 \hat{\alpha}_i / \hat{\sigma}_i^2 = \dot{\beta}$, $\hat{\delta}_{\hat{\imath}i}^2 \hat{\alpha}_i / \hat{\sigma}_i^2 = \min \left(\hat{\delta}_{\hat{\imath}i}^2 \hat{\alpha}_i / \hat{\sigma}_i^2, \right.$ $\left. \hat{\delta}_{i\hat{\imath}}^2 \hat{\alpha}_i / \hat{\sigma}_i^2 \right)$ must hold. Since $\hat{\delta}_{\hat{\imath}i}^2 < \hat{\delta}_{i\hat{\imath}}^2$ is obviously equivalent to $\hat{\delta}_{\hat{\imath}i}^2 \hat{\alpha}_i / \hat{\sigma}_i^2 = \min$ $\left(\hat{\delta}_{\hat{\imath}i}^2 \hat{\alpha}_i / \hat{\sigma}_i^2, \hat{\delta}_{i\hat{\imath}}^2 \hat{\alpha}_i / \hat{\sigma}_i^2 \right)$, $\hat{\delta}_{\hat{\imath}i}^2 \hat{\alpha}_i / \hat{\sigma}_i^2 = \dot{\beta}$ leads to $\hat{\delta}_{\hat{\imath}i}^2 < \hat{\delta}_{i\hat{\imath}}^2$.

Thus, (5.38) can be further simplified as

$$
\frac{\mathbb{N}_i^2/\hat{\sigma}_i^2}{\mathbb{N}_{\check{\imath}}^2/\hat{\sigma}_{\check{\imath}}^2} = \frac{\sum_{i \in \left\{ i \mid \hat{\delta}_{\hat{\imath}i}^2 < \hat{\delta}_{i\hat{\imath}}^2 \right\}} 1/\hat{\delta}_{\hat{\imath}i}^2}{\sum_{j \in \left\{ j \mid \hat{\delta}_{\check{\imath}j}^2 > \hat{\delta}_{j\check{\imath}}^2 \right\}} 1/\hat{\delta}_{j\check{\imath}}^2}. \tag{5.39}
$$

From (5.39), we can see that any design $i \notin \{\hat{\imath}, \check{\imath}\}$ can serve as an object of reference. Its relative distances $\hat{\delta}_{\hat{\imath}i}^2$ and $\hat{\delta}_{j\check{\imath}}^2$ determine the allocation ratio of budgets between $\hat{\imath}$ and $\check{\imath}$. However, only the smaller one of $\hat{\delta}_{\hat{\imath}i}^2$ and $\hat{\delta}_{i\hat{\imath}}^2$ is considered in (5.39) due to the simplification of the derived allocation rules from (5.36) and (5.37), which do not consider the following two extreme cases:

Case 2: If

$$
\hat{\delta}_{\hat{\imath}i}^2 > \hat{\delta}_{i\hat{\imath}}^2, \quad \forall i \notin \{\hat{\imath}, \check{\imath}\}. \tag{5.40}
$$

Case 3: If

$$
\hat{\delta}_{\hat{\imath}i}^2 < \hat{\delta}_{i\hat{\imath}}^2, \quad \forall i \notin \{\hat{\imath}, \check{\imath}\}. \tag{5.41}
$$

Figures 5.1 and 5.2 illustrate the above two cases, respectively. In them, $\sum_{i \in \left\{ i \mid \hat{\delta}_{\hat{\imath}i}^2 < \hat{\delta}_{i\hat{\imath}}^2 \right\}} 1/\hat{\delta}_{\hat{\imath}i}^2 = 0$ or $\sum_{j \in \left\{ j \mid \hat{\delta}_{\check{\imath}j}^2 > \hat{\delta}_{j\check{\imath}}^2 \right\}} 1/\hat{\delta}_{j\check{\imath}}^2 = 0$ in (5.39), which means that the

Figure 5.1 All designs except $\hat{i}$ and $\check{i}$ approximate $\bar{i}$.

Figure 5.2 All designs except $\hat{i}$ and $\check{i}$ approximate $\hat{i}$.

comparison between $\hat{i}$ and $\check{i}$ is lost such that the allocation ratio of budgets between $\hat{i}$ and $\check{i}$ cannot be obtained.

We adopt an intuitive allocation rule as follows in order to deal with these two special cases: if

$$\hat{\delta}_{\hat{i}i}^2 > \hat{\delta}_{\check{i}i}^2, \quad \forall i \notin \{\hat{i}, \check{i}\},$$

then

$$\frac{\mathbb{N}_{\hat{i}}^2/\hat{\sigma}_{\hat{i}}^2}{\mathbb{N}_{\check{i}}^2/\hat{\sigma}_{\check{i}}^2} = \frac{1/\min_{i \notin \{\hat{i}, \check{i}\}} \hat{\delta}_{\check{i}i}^2}{\sum_{j \in \left\{ j | \hat{\delta}_{\hat{i}j}^2 > \hat{\delta}_{\check{i}j}^2 \right\}} 1/\hat{\delta}_{\check{j}i}^2}, \quad \forall j \notin \{\hat{i}, \check{i}\}. \tag{5.42}$$

When $\hat{i}$ is a target and no design $i \notin \{\hat{i}, \check{i}\}$ exists satisfying $\hat{\delta}_{\hat{i}i}^2 < \hat{\delta}_{\check{i}i}^2$, the one with the smallest distance $\hat{\delta}_{\check{i}i}^2$ is selected as the object of reference for $\hat{i}$ relative to $\check{i}$. In this way, the allocation ratio of budgets between $\hat{i}$ and $\check{i}$ is obtained as in (5.42). Similarly, if

$$\hat{\delta}_{\hat{i}i}^2 < \hat{\delta}_{\check{i}i}^2, \quad \forall i \notin \{\hat{i}, \check{i}\},$$

then

$$\frac{\mathbb{N}_{\hat{i}}^2/\hat{\sigma}_{\hat{i}}^2}{\mathbb{N}_{\check{i}}^2/\hat{\sigma}_{\check{i}}^2} = \frac{\sum_{i \in \left\{ i | \hat{\delta}_{\hat{i}i}^2 < \hat{\delta}_{\check{i}i}^2 \right\}} 1/\hat{\delta}_{\hat{i}i}^2}{1/\min_{j \notin \{\hat{i}, \check{i}\}} \hat{\delta}_{\check{j}i}^2}, \quad \forall i \notin \{\hat{i}, \check{i}\}. \tag{5.43}$$

We define $\dot{\Omega}_1 = \left\{ i | \hat{\delta}_{\hat{i}i}^2 < \hat{\delta}_{\check{i}i}^2, \forall i \notin \{\hat{i}, \check{i}\} \right\}$, $\dot{\Omega}_2 = \left\{ i | \min_{i \notin \{\hat{i}, \check{i}\}} \hat{\delta}_{\hat{i}i}^2 \right\}$, $\Phi_1 = \left\{ j | \hat{\delta}_{\hat{i}j}^2 > \hat{\delta}_{\check{j}i}^2, \forall j \notin \{\hat{i}, \check{i}\} \right\}$ and $\Phi_2 = \left\{ j | \min_{j \notin \{\hat{i}, \check{i}\}} \hat{\delta}_{\check{j}i}^2 \right\}$, respectively. Thus, according to (5.39), (5.42), and (5.43), we conclude the relationship between $\mathbb{N}_{\hat{i}}$ and $\mathbb{N}_{\check{i}}$ as

$$\frac{\mathbb{N}_{\hat{i}}^2/\hat{\sigma}_{\hat{i}}^2}{\mathbb{N}_{\check{i}}^2/\hat{\sigma}_{\check{i}}^2} = \frac{\sum_{i \in \{\dot{\Omega}_1 \bigcup \dot{\Omega}_2\}} 1/\hat{\delta}_{\hat{i}i}^2}{\sum_{j \in \{\Phi_1 \bigcup \Phi_2\}} 1/\hat{\delta}_{\check{j}i}^2}, \tag{5.44}$$

which can now deal with all the cases.

According to (5.25), (5.30), and (5.44), we have the following main result.

Theorem 5.1 *Given a total number of simulation samples $\bar{T}$ are allocated to $\hat{k}$ competing designs, whose performances are depicted by random variables with means J_i and variances $\hat{\sigma}_i^2$. As $\bar{T} \to +\infty$, APCS can be asymptotically when*

$$(R1) \quad \frac{\mathbb{N}_{\hat{i}}^2}{\hat{\sigma}_{\hat{i}}^2} + \frac{\mathbb{N}_{\check{i}}^2}{\hat{\sigma}_{\check{i}}^2} = \sum_{i \neq \hat{i}, \check{i}} \frac{\mathbb{N}_i^2}{\hat{\sigma}_i^2},$$

$$(R2) \quad \frac{\mathbb{N}_i}{\mathbb{N}_j} = \frac{\hat{\alpha}_i}{\hat{\alpha}_j} = \frac{\hat{\sigma}_i^2 / \min\left(\hat{\delta}_{\hat{i}i}^2, \hat{\delta}_{\check{i}i}^2\right)}{\hat{\sigma}_j^2 / \min\left(\hat{\delta}_{\hat{i}j}^2, \hat{\delta}_{ji}^2\right)}, \quad and$$

$$(R3) \quad \frac{\mathbb{N}_{\hat{i}}^2 / \hat{\sigma}_{\hat{i}}^2}{\mathbb{N}_{\check{i}}^2 / \hat{\sigma}_{\check{i}}^2} = \frac{\sum_{i \in \{\Omega_1 \cup \Omega_2\}} 1 / \hat{\delta}_{\hat{i}i}^2}{\sum_{j \in \{\Phi_1 \cup \Phi_2\}} 1 / \hat{\delta}_{ji}^2}.$$

∎

To approach the effect of Theorem 5.1, this section presents a sequential approach to select the best and worst designs simultaneously out of $\hat{k}$ alternatives with a given computing budget.

Initially, $\bar{n}_0$ simulation replications for each design are conducted. As simulation proceeds, the sample mean as well as the sample variance of each design are computed at each stage, where the sample variance is used to approximate $\hat{\sigma}_i^2$. According to this collected simulation output, an incremental computing budget Δ is allocated to designs based on Theorem 5.1 at each stage. This procedure continues until the total budget $\bar{T}$ is exhausted. It is realized in Algorithm OCBAbw.

OCBAbw Algorithm

Input: Number of simulation replications for each design $\bar{n}_0$, number of designs $\hat{k}$, set of designs Θ, incremental computing budget Δ, total number of simulation replications $\bar{T}$.

Output: The best design.

1: Perform $\bar{n}_0$ simulation replications for all designs;

$\quad \bar{l} \leftarrow 0$;

$\quad \mathbb{N}_{1,\bar{l}} = \mathbb{N}_{2,\bar{l}} = \cdots = \mathbb{N}_{\hat{k},\bar{l}} = \bar{n}_0$.

2: If $\sum_{i \in \Theta} \mathbb{N}_{i,\bar{l}} + \Delta > \bar{T}$ stop.

3: Increase the computing budget by Δ;

Use Theorem 5.1 to compute the new budget allocation $\mathbb{N}_{1,\bar{l}+1}, \mathbb{N}_{2,\bar{l}+1}, \ldots, \mathbb{N}_{\hat{k},\bar{l}+1}$.

4: Simulate design i for additional max $\left(0, \mathbb{N}_{i,\bar{l}+1} - \mathbb{N}_{i,\bar{l}}\right)$ replications, $i \in \Theta$;

$\bar{l} \leftarrow \bar{l} + 1$.

Go to step 2.

END

In Algorithm OCBAbw, l is the iteration count. As simulation evolves, designs $\hat{i}$ and $\check{i}$, which are the designs with the smallest and the largest sample means, respectively, may change from iteration to iteration. Designs $\hat{i}$ and $\check{i}$ converge to the best and worst designs when $\mathbb{N}_{\hat{i}} \to +\infty$ and $\mathbb{N}_{\check{i}} \to +\infty$. When the changes happen, the previous design $\hat{i}$ may not be simulated at all in the next several iterations according to $\max(0, \mathbb{N}_{i,\bar{l}+1} - \mathbb{N}_{i,\bar{l}})$, because it may already have enough accumulated simulation replications in previous iterations. $\mathbb{N}_i$ is an integer instead of a real number from Theorem 5.1 when conducting tests. This change has little influence on testing our algorithm because we adhere to two principles. First, the number of simulation replications allocated to each design approximates to the allocation sequence from Theorem 5.1 as closely as possible. Second, we must ensure that the sum of simulation replications newly allocated to each design equals a fixed value ($\it{\Delta}$) in each iteration. Especially in the last iteration, if the value of $\mathbb{N}_i$ is less than the previous value, no replications are allocated to $\mathbb{N}_i$. In order to prevent other designs from adding more replications than the rest of replications, all designs except $\mathbb{N}_i$ share replications under $\it{\Delta}$ that is limit to $(\bar{T} - \sum_{i \in \Theta} \mathbb{N}_{i,\bar{l}})$. Thus, the total replications allocated to all designs is no more than the computing budget.

After considering selecting the best and worst designs equally important in the above discussion, we now discuss a case in which selecting the best design and the worst one with different weights.

First, we define the weight of selecting the best design as $\bar{r}_1 > 0$ and the weight of selecting the worst one as $\bar{r}_2 > 0$, $\bar{r}_1 + \bar{r}_2 = 2$. Neither $\bar{r}_1$ nor $\bar{r}_2$ could be 0 so as to both $\hat{i}$ and $\check{i}$ should be considered. Obviously, when $\bar{r}_1 = \bar{r}_2 = 1$, it means that the selections of the best and worst designs are equally important. When $\bar{r}_1 > \bar{r}_2$ or $\bar{r}_2 > \bar{r}_1$, it means that selecting the best design is more or less important than the worst one, respectively. Similar to $P\{CS\}$, $P\{CS_r\}$ is defined as the probability to correctly select the best design and worst design with respective weights $\bar{r}_1$ and $\bar{r}_2$. Thus, a new lower bound $APCS_r$ for $P\{CS_r\}$ is

$$xP\{CS_r\} \geq 1 - \bar{r}_1 \cdot \sum_{i \neq \hat{i}, \check{i}} P\{\tilde{J}_{\hat{i}} > \tilde{J}_i\} - \bar{r}_2 \cdot \sum_{i \neq \hat{i}, \check{i}} P\{\tilde{J}_i > \tilde{J}_{\check{i}}\}$$

$$= APCS_r. \tag{5.45}$$

The whole proof is similar to OCBAbw and thus is omitted. The new rules among $\mathbb{N}_{\hat{i}}$, $\mathbb{N}_{\check{i}}$, and $\mathbb{N}_i$ with weights in selecting the best and worst designs is presented in Theorem 5.2.

Theorem 5.2 *Let two weights $\bar{r}_1 > 0$ and $\bar{r}_2 > 0$ represent the importance of selecting the best and worst designs, respectively, $\bar{r}_1 + \bar{r}_2 = 2$. A total number of simulation samples $\bar{T}$ are to be allocated to $\hat{k}$ competing designs. The performances of these designs are depicted by random variables with means J_i and variances $\hat{\sigma}_i^2$.*

As $\bar{T} \to +\infty$, $APCS_r$ can be asymptotic if

$$(R4) \quad \frac{\aleph_{\hat{i}}^2}{\hat{\sigma}_{\hat{i}}^2} + \frac{\aleph_{\check{i}}^2}{\hat{\sigma}_{\check{i}}^2} = \sum_{\substack{i \neq \hat{i} \\ i \neq \check{i}}} \frac{\aleph_i^2}{\hat{\sigma}_i^2},$$

$$(R5) \quad \frac{\aleph_i}{\aleph_j} = \frac{\hat{\alpha}_i}{\hat{\alpha}_j} = \frac{\hat{\sigma}_i^2/\min\left(\hat{\delta}_{\hat{i}i}^2, \hat{\delta}_{\check{i}i}^2\right)}{\hat{\sigma}_j^2/\min\left(\hat{\delta}_{\hat{i}j}^2, \hat{\delta}_{ji}^2\right)}, \quad and$$

$$(R6) \quad \frac{\aleph_{\hat{i}}^2/\hat{\sigma}_{\hat{i}}^2}{\aleph_{\check{i}}^2/\hat{\sigma}_{\check{i}}^2} = \frac{\bar{r}_1 \cdot \sum_{i \in \{\Omega_1 \cup \Omega_2\}} 1/\hat{\delta}_{\hat{i}i}^2}{\bar{r}_2 \cdot \sum_{j \in \{\Phi_1 \cup \Phi_2\}} 1/\hat{\delta}_{ji}^2},$$

where $\dot{\Omega}_1 = \left\{i | \hat{\delta}_{\hat{i}i}^2 < \hat{\delta}_{\check{i}i}^2, \forall i \neq \hat{i}, \check{i}\right\}$, $\Omega_2 = \left\{i | \min_{i \neq \hat{i}, \check{i}} \hat{\delta}_{\hat{i}i}^2\right\}$,
$\Phi_1 = \left\{j | \hat{\delta}_{\hat{i}j}^2 > \hat{\delta}_{ji}^2, \forall j \neq \hat{i}, \check{i}\right\}$ *and* $\Phi_2 = \left\{j | \min_{j \neq \hat{i}, \check{i}} \hat{\delta}_{ji}^2\right\}$. ∎

5.2.3 Simulation Studies

It is necessary to test the introduced algorithm and compare it with several different allocation procedures. This section conducts numerical tests in different environments.

Equal allocation is a naive way to select the target design(s) that the simulation replications are allocated equally to each design. Namely, the number of simulation replications of each design is $\bar{T}/\hat{k}$. According to [2, 17], the performance of equal allocation can serve as a benchmark for comparison.

Proportional to variance (PTV) procedure sequentially determines, at every iteration, the number of simulation replications for each design based on the newly updated sample variance. First of all, all designs are simulated for $\bar{n}_0$ replications, and calculate the sample mean, as well as the sample variance. At every iteration, the ratios among all designs are calculated as the following continued equality:

$$\frac{\aleph_1^{\bar{l}+1}}{\hat{s}_1^2} = \frac{\aleph_2^{\bar{l}+1}}{\hat{s}_2^2} = \cdots = \frac{\aleph_k^{\bar{l}+1}}{\hat{s}_k^2},$$

where $\sum \aleph_i^{\bar{l}+1} = \sum \aleph_i^{\bar{l}} + \dot{\Delta}$. Thus, design i is conducted $\max\left(0, \aleph_i^{\bar{l}+1} - \aleph_i^{\bar{l}}\right)$ replications.

Half Budget for the Best and Half Budget for the Worst using OCBA (HalfT) selects either of best and worst designs by spending a half budget. The procedure is inspired and derived from the original OCBA based on the fact that the original OCBA, which considers only the best design, is not suitable for selecting the best and worst designs. In detail, it firstly assigns half budget $\bar{T}/2$ to select the best design using OCBA. Secondly, it assigns the other half budget $\bar{T}/2$ to select the worst design using OCBA. In the second stage, when it selects the worst design using OCBA, it regards the worst design as the "best" design. It is worth noting that we retain the sample information for selecting the best design (or the worst one) in the first stage. When selecting the worst design (or the best one) in the second stage, we use this information.

OCBA is a sequential procedure for selecting the best design. Its efficiency has been demonstrated in many fields. However, it is not suitable for the problem that needs to select both the best and worst designs. Thus, we propose an heuristic algorithm named OCBAht which is able to select both the best and worst designs simultaneously. If only the worst design is considered, it is the same as only considering selecting the best design. We can get the allocations for each design by directly applying OCBA in each situation. Then one intuitive idea borrowed from OCBA of selecting the best and worst designs together is to take an average on the allocations given by these two situations.

This section conducts these five procedures in four environments. In each environment, we compare the performances of the convergence of these five algorithms. Meanwhile, numerical evidences on the rate of convergence of the algorithms are given in tables to show the value of $P\{CS\}$ per 500 replications. We use "Equal" to stand for Equal Allocation and "HalfT" for the procedure using half budget for the best design and another half for the worst one. Except "Equal," each procedure needs an initial budget $\bar{n}_0$ for each design and an extra budget $\acute{\Delta}$ in every iteration while "HalfT" needs them twice. We set $\bar{n}_0 = 20$ and $\acute{\Delta} = 50$.

Environment 5.1 (Equal Variance): First, there are 10 alterative designs, with distribution $N\left(i, 6^2\right)$ for design $i = 1, 2, \ldots, 10$. Each design has the same variance. The target is to select the best and worst designs correctly, i.e., designs 1 (the best design) and 10 (the worst one). From Fig. 5.3, it can be seen that all procedures obtain higher $P\{CS\}$ as the available computing budget increases. Among all procedures, OCBAbw achieves the fastest speed to reach high $P\{CS\}$ with a lower amount of computing budget than other procedures. In order to show the numerical evidence on the rate of convergence of these five algorithms, we collect the values of $P\{CS\}$ per 500 replications in Table 5.1. In this experiment, it is shown from Table 5.2 that when $P\{CS\}$ of OCBAbw reaches 99%, that of Equal is 89.79%, HalfT 98.32%, OCBAht 98.85%, and PTV 90.48%. The numeric results reveal that OCBAbw outperforms other procedures.

A larger variance for each design is necessary to test the performance of these five procedures. We again use 10 alterative designs, but with distribution $N\left(i, 2 \times 6^2\right)$ for design $i = 1, 2, \ldots, 10$. Each design has the same large variance. From Fig. 5.4, it can be seen that all procedures obtain higher $P\{CS\}$ as the

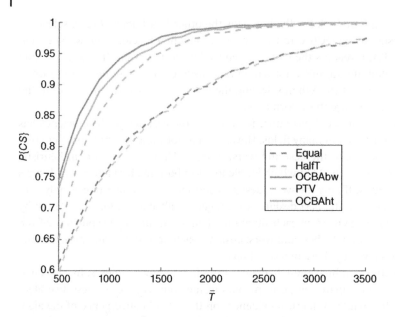

Figure 5.3 $P\{CS\}$ versus $\bar{T}$ when five different allocation procedures are used for Experiment 1 in Environment 5.1.

Table 5.1 Numerical evidence on the rate of convergence in Environment 5.1 about the value of $P\{CS\}$ per 500 replications.

Replication	Equal	HalfT	OCBAbw	OCBAht	PTV
500	0.611400	0.651500	0.748000	0.732500	0.595100
1000	0.767900	0.876600	0.925000	0.906900	0.764800
1500	0.853600	0.951200	0.976800	0.967300	0.854600
2000	0.897900	0.983200	0.990500	0.988500	0.904800
2500	0.937700	0.992200	0.996500	0.994600	0.936700
3000	0.958800	0.995700	0.998500	0.998500	0.957600
3500	0.974100	0.998500	0.999600	0.999300	0.973800

available computing budget increases. Among all procedures, OCBAbw is similarly the fastest to reach high $P\{CS\}$ with a lower amount of computing budget. The values of $P\{CS\}$ per 500 replications collected in Table 5.3 verify the performance of the convergence of these five algorithms. In this experiment, it is shown from Table 5.2 that when $P\{CS\}$ of OCBAbw reaches 99%, that of Equal

Table 5.2 $P\{CS\}$ of other procedures when OCBAbw reaches 99% in Environment 5.1.

Procedure	Experiment 1 (%)	Experiment 2 (%)
Equal	89.79	91.29
HalfT	98.32	98.04
OCBAht	98.85	98.87
PTV	90.48	91.22

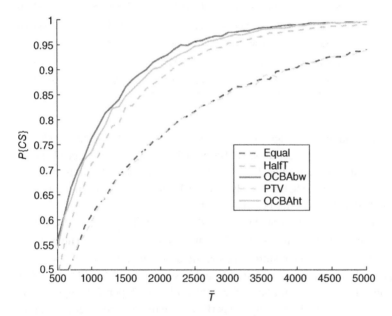

Figure 5.4 $P\{CS\}$ versus $\bar{T}$ when five different allocation procedures are used for Experiment 2 in Environment 5.1.

is 91.29%, HalfT 98.04%, OCBAht 98.87%, and PTV 91.22%. The numeric results reveal that OCBAbw is superior to other procedures.

Environment 5.2 (Flat and Steep Cases): It is desired to test the performance of the proposed procedure under a kind of environment where the interval between two adjacent designs is not equal. Under such environment, there are two cases: flat and steep cases, which are illustrated in [2].

Firstly, the performance of the proposed approach in the flat case is tested. There are 10 alterative designs, with distribution $N(10 - 3\sqrt{10 - i}, 6^2)$ for design

Table 5.3 Numerical evidence on the rate of convergence in Environment 5.2 about the value of $P\{CS\}$ per 500 replications.

Replication	Equal	HalfT	OCBAbw	OCBAht	PTV
500	0.437800	0.487400	0.560300	0.543800	0.425200
1000	0.609000	0.711200	0.761800	0.734800	0.597000
1500	0.703600	0.827000	0.865700	0.847600	0.693600
2000	0.763700	0.881900	0.922800	0.904700	0.764300
2500	0.816400	0.925200	0.956100	0.946700	0.814600
3000	0.856700	0.953500	0.974400	0.966800	0.852100
3500	0.881300	0.966600	0.984200	0.981000	0.886700
4000	0.903900	0.978000	0.989000	0.988700	0.903000
4500	0.926400	0.987300	0.994200	0.994300	0.925900
5000	0.939300	0.990300	0.995900	0.995300	0.938100

$i = 1, 2, \ldots, 10$ and each design has the same variance. The result is shown in Fig. 5.5. Among all procedures, OCBAbw achieves the fastest speed. From Table 5.4, it is shown that when $P\{CS\}$ of OCBAbw reaches 99%, that of Equal is 89.18%, HalfT 94.35%, OCBAht 97.49%, and PTV 88.62%. The numeric results reveal that OCBAbw has better performance than the others.

Second, these methods are tested with the steep case, still among 10 alternative designs. Design $i = 1, 2, \ldots, 10$ has distribution $N\left(1 + 3\sqrt{i-1}, 6^2\right)$. The result is shown in Fig. 5.6. The same results as before are observed. From Table 5.4, when $P\{CS\}$ of OCBAbw reaches 99%, that of Equal is 89.17%, HalfT 96.67%, OCBAht 97.57%, and PTV 88.23%. Similarly, the values of $P\{CS\}$ per 500 replications collected in Tables 5.5 and 5.6 verify the performance of the convergence of these five algorithms.

Environment 5.3 (Unequal Variance): Now consider a kind of environment where the variances of the distribution are not equal. First, the performance is tested in the environment where the variances of designs increase with mean: design $i = 1, 2, \ldots, 10$ follow distribution $N(i, i)$. Apparently, the best design has the smallest variance and the worst one has the largest variance.

From Table 5.7, when $P\{CS\}$ of OCBAbw reaches 99%, that of Equal is 87.05%, HalfT 97.23%, OCBAht 97.45%, and PTV 94.36%. OCBAbw clearly outperforms other procedures as well.

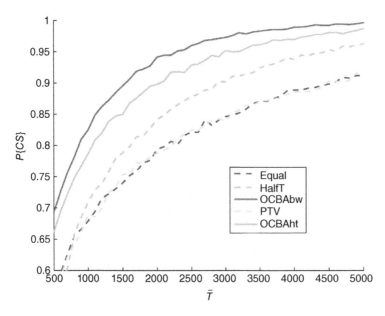

Figure 5.5 $P\{CS\}$ versus $\bar{T}$ when five different allocation procedures are used for Experiment 1 in Environment 5.2.

Table 5.4 $P\{CS\}$ of other procedures when OCBAbw reaches 99% in Environment 5.2.

Procedure	Experiment 1 (%)	Experiment 2 (%)
Equal	89.18	89.17
HalfT	94.35	96.67
OCBAht	97.49	97.57
PTV	88.62	88.23

Second, the performance is tested in the environment where the design's variances decrease with their means: design $i = 1, 2, \ldots, 10$ follow distribution $N(i, 11 - i)$. Oppositely, the best design has the largest variance and the worst one has the smallest variance. From Table 5.7, when $P\{CS\}$ of OCBAbw reaches 99%, that of Equal is 87.18%, HalfT 93.11%, OCBAht 97.14%, and PTV 94.36%. Figures 5.7 and 5.8 show the results of the experiments of Environment 5.3 and the values of $P\{CS\}$ per 500 replications collected in Tables 5.8 and 5.9 verify the performance of the convergence of these five algorithms.

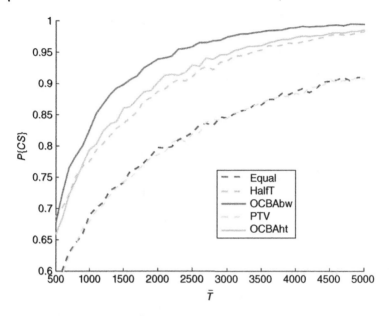

Figure 5.6 $P\{CS\}$ versus $\bar{T}$ when five different allocation procedures are used for Experiment 2 in Environment 5.2.

Table 5.5 Numerical evidence on the rate of convergence in Environment 5.3 about the value of $P\{CS\}$ per 500 replications.

Replication	Equal	HalfT	OCBAbw	OCBAht	PTV
500	0.572000	0.523800	0.693500	0.662200	0.559700
1000	0.679300	0.709500	0.824700	0.786900	0.690100
1500	0.739800	0.788400	0.895200	0.849800	0.753200
2000	0.794200	0.841200	0.941200	0.897900	0.788200
2500	0.821200	0.878400	0.959800	0.928700	0.825100
3000	0.846300	0.901900	0.972800	0.951800	0.840300
3500	0.870700	0.924200	0.982900	0.961200	0.870000
4000	0.888800	0.937700	0.989300	0.969500	0.885300
4500	0.898500	0.953600	0.992200	0.980700	0.902500
5000	0.911500	0.962900	0.996200	0.986800	0.916500

Table 5.6 Numerical evidence on the rate of convergence in Environment 5.4 about the value of $P\{CS\}$ per 500 replications.

Replication	Equal	HalfT	OCBAbw	OCBAht	PTV
500	0.575300	0.687200	0.678100	0.659500	0.568800
1000	0.689500	0.774800	0.825700	0.794300	0.679300
1500	0.740900	0.835800	0.898000	0.860300	0.742000
2000	0.797600	0.887000	0.938700	0.901800	0.786100
2500	0.821200	0.910600	0.958500	0.929300	0.818600
3000	0.844800	0.935700	0.974000	0.946000	0.844100
3500	0.865500	0.953300	0.982500	0.960300	0.868700
4000	0.887700	0.963500	0.989000	0.973700	0.884200
4500	0.903400	0.976700	0.993000	0.980000	0.906300
5000	0.910400	0.982700	0.994600	0.985800	0.907700

Table 5.7 $P\{CS\}$ of other procedures when OCBAbw reaches 99% in Environment 5.3.

Procedure	Experiment 1 (%)	Experiment 2 (%)
Equal	87.05	87.18
HalfT	97.23	93.11
OCBAht	97.45	97.14
PTV	94.36	94.36

Environment 5.4 (Larger-Scale Problem): Let us examine their performances in a larger-scale problem. Distinct from the above three environments, Environment 5.4 has 50 designs with distribution $N(i, 10)$ for design $i = 1, 2, \ldots, 50$. Figure 5.9 shows the result of Environment 5.4. Among all procedures, the gap between Equal, PTV, and others is large, and they are both inferior to others. OCBAbw reaches high $P\{CS\}$ faster than other procedures as well, that is supported by Table 5.10.

From the above figures and tables, we conclude that the OCBAbw derived from the original OCBA significantly enhances the efficiency to solve the problem. Although both HalfT and OCBAht are modified versions of OCBA, their efficiency is not so great in comparison with OCBAbw. HalfT assigns the half budget for the best design and the other half for the worst one, the budgets for selecting them

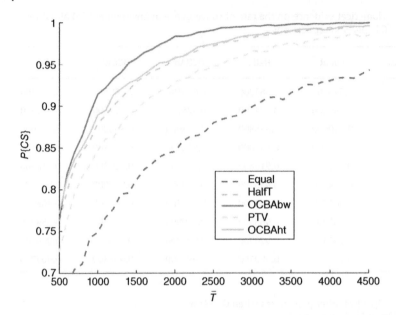

Figure 5.7 $P\{CS\}$ versus $\bar{T}$ when five different allocation procedures are used for Experiment 1 in Environment 5.3.

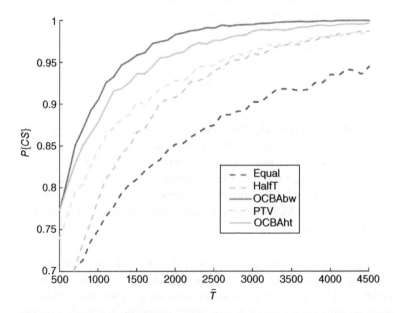

Figure 5.8 $P\{CS\}$ versus $\bar{T}$ when five different allocation procedures are used for Experiment 2 in Environment 5.3.

Table 5.8 Numerical evidence on the rate of convergence in Environment 5.5 about the value of $P\{CS\}$ per 500 replications.

Replication	Equal	HalfT	OCBAbw	OCBAht	PTV
500	0.653200	0.763500	0.765200	0.772400	0.726300
1000	0.748800	0.879800	0.914300	0.889700	0.847000
1500	0.811200	0.929200	0.956900	0.931900	0.894900
2000	0.845400	0.957900	0.983800	0.958500	0.929400
2500	0.880000	0.972300	0.991100	0.974500	0.948400
3000	0.899200	0.981600	0.996200	0.985100	0.966300
3500	0.916000	0.989500	0.998300	0.990000	0.976200
4000	0.930500	0.994300	0.998900	0.991800	0.982700
4500	0.943000	0.994700	1.000000	0.996500	0.985000

Table 5.9 Numerical evidence on the rate of convergence in Environment 5.6 about the value of $P\{CS\}$ per 500 replications.

Replication	Equal	HalfT	OCBAbw	OCBAht	PTV
500	0.646800	0.601500	0.772300	0.772900	0.739300
1000	0.748900	0.786100	0.905400	0.878800	0.845900
1500	0.810000	0.866300	0.958400	0.956400	0.902200
2000	0.851800	0.908000	0.982800	0.960200	0.928300
2500	0.874500	0.940000	0.990300	0.975900	0.946400
3000	0.902100	0.960400	0.994600	0.986600	0.964600
3500	0.917200	0.971100	0.998000	0.989300	0.973800
4000	0.935200	0.979100	0.998800	0.992900	0.981800
4500	0.944500	0.986600	0.999500	0.996600	0.987400

are fixed instead of adaptive allocation for the best and the worst as in OCBAbw. As for OCBAht, its allocation sequence needs to be computed twice instead of once in each iteration. OCBAht approaches closely to OCBAbw compared with other algorithms. In summary, the OCBAbw is better than the other procedures from numerical results.

The philosophy of selecting the best and worst has been applied to many fields, especially to performance appraisal of employment in a company [27]. We call it Winner Up Loser Out (WULO), which is one of the widely used incentive schemes.

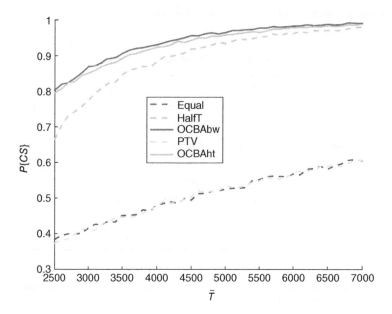

Figure 5.9 $P\{CS\}$ versus $\bar{T}$ when five different allocation procedures are used for the experiment in Environment 5.4.

Table 5.10 Numerical evidence on the rate of convergence in Environment 5.7 about the value of $P\{CS\}$ per 500 replications.

Replication	Equal	HalfT	OCBAbw	OCBAht	PTV
2500	0.383900	0.665500	0.802800	0.796700	0.372800
3000	0.415400	0.774500	0.869000	0.851200	0.415000
3500	0.451000	0.846500	0.905600	0.892200	0.447600
4000	0.480600	0.883400	0.930200	0.923700	0.476400
4500	0.494400	0.918800	0.956300	0.943100	0.504400
5000	0.524700	0.934800	0.964700	0.958000	0.535500
5500	0.551600	0.952700	0.977700	0.965200	0.547800
6000	0.567400	0.962600	0.983000	0.978900	0.571800
6500	0.580900	0.970500	0.988000	0.984100	0.589600
7000	0.604800	0.980600	0.990400	0.988400	0.606300

It describes the situation where the prior performance of the employments affects their future success [28]. Winner up and loser out are two independent strategies that a company usually adopts. The strategy of winner up contributes to the aggressiveness in the company, but it may disrupt the maintenance. In contrast, the strategy of loser out results in the minimal aggressiveness, but it encourages the stability of the company [29]. Therefore, the strategy of the combination of winner up and loser out is usually more adapted to companies for more benefit. Especially in [30], winner up and loser out are combined to be applied in a hospital and show better performance in human resource management than using them alone. In the company, the employees are usually ranked according to their performance appraisal. Then the first one is rewarded and the last one is penalized. However, such strategy usually does not alter an individual's actual ability, but rather affects his/her future performance or aggressiveness, thereby eventually affecting the performance of the whole company. Thus, whether the strategy of WULO increases the benefits of a company will be explored to show the motivation and the potential application.

Example 5.3 Assume a company with one principal and $\bar{\kappa}$ employees. All outcome of this company is owned by the principal, but the principal should pay all employees salary for their contribution. We assume the only way to know the contribution of each employee for the principal is his/her performance appraisal. Thus, the strategy of WULO usually serves as an effective strategy for the principal to stimulate the employees to work hard for more outcome. For employee $i \in \{1, 2, \ldots, \bar{\kappa}\}$, we use $\hat{d}_i$ to represent his/her outcome per unit of salary, which is affected by many issues, i.e., actual ability, motivation, and loyalty. Thus, $\hat{d}_i$ serves as the standard to appraise the performance of the employment. We define $\hat{w}_i$ and $\hat{p}_i$ as the salary and outcome of employee i, respectively. Meanwhile, the salary total for all employees is a constant number $\hat{W}$, i.e., $\sum_{i=1}^{\bar{\kappa}} \hat{w}_i = \hat{W}$. Thus, the outcome of employee i can be expressed as $\hat{p}_i = \hat{w}_i * \hat{d}_i$. How to maximize total company outcome $\hat{P}$ is of interest to the principal. Mathematically,

$$\max \hat{P} = \sum_{i=1}^{\bar{\kappa}} \hat{p}_i,$$

$$\text{s.t.} \sum_{i \in \{1,2,\ldots,\bar{\kappa}\}} \hat{w}_i = \hat{W} \quad \text{and} \quad \hat{w}_i \geq 0, \tag{5.46}$$

where $\hat{p}_i = \hat{w}_i * \hat{d}_i$.

In order to show the efficiency of the strategy of WULO, we test it in the above Environment 5.1. That is to say, the number of employees $\bar{\kappa}$ is set as 10 and the outcome per unit of salary of employee i is i, i.e., $\hat{d}_i = i, i \in \{1, 2, \ldots, 10\}$. $\bar{T}$ represents the expense of one time performance appraisal, set as 1000 in the experiment.

Table 5.11 The percentage of the improvement of OCBAbw over the other four procedures.

$\hat{R}_{OCBAht}$	$\hat{R}_{HalfT}$	$\hat{R}_{PTV}$	$\hat{R}_{Equal}$
0.2%	0.5%	2.1%	2.1%

An intuitive strategy of WULO is adopted. Specifically, the salary of employee $\check{i}$ who ranks the last according to his performance appraisal is given to employee $\hat{i}$ who ranks the first. Meanwhile, we let the salary of each employee be a constant $\hat{w}_0 = \hat{W}/\bar{k}$ for simplification, i.e., $\hat{w}_i = \hat{w}_0, i \in \{1, 2, \ldots, 10\}$. Thus, the total company outcome $\hat{P}'$ after adopting the strategy of WULO is $\hat{P}' = \sum_{i=1, i \notin \{\hat{i}, \check{i}\}}^{10} \hat{w}_0 \hat{d}_i + 2\hat{w}_0 \hat{d}_{\hat{i}} + (\hat{w}_0 - \hat{w}_0)\hat{d}_{\check{i}}$ and the efficiency of the strategy of WULO can be expressed as

$$\max(\hat{P}' - \hat{P})$$

$$= \sum_{i=1, i \notin \{\hat{i}, \check{i}\}}^{10} \hat{w}_0 \hat{d}_i + 2\hat{w}_0 \hat{d}_{\hat{i}} + (\hat{w}_0 - \hat{w}_0)\hat{d}_{\check{i}} - \sum_{i=1}^{10} \hat{w}_0 \hat{d}_i$$

$$= \hat{w}_0 (\hat{d}_{\hat{i}} - \hat{d}_{\check{i}}). \tag{5.47}$$

From (5.47), $\hat{i}$ and $\check{i}$ are the best and worst employees with the maximum and minimum $\hat{d}_i$ and the improvement in efficiency when adopting WULO is related closely to the accuracy of convergence to the true best and the true worst employees. That is to say, the higher the accuracy of selecting the best and worst employees, the more improvement in efficiency of WULO. Thus, the target of (5.47) is same to the one of (5.10). Similarly, the improvement in efficiency of OCBAbw is compared with OCBAht, HalfT, PTV, and Equal in Environment 5.1. If we use $\hat{R}_{OCBAbw}, \hat{R}_{OCBAht}, \hat{R}_{HalfT}, \hat{R}_{PTV}$, and $\hat{R}_{Equal}$ to represent the results of these five procedures, respectively, then the percentage of the improvement of OCBAbw over the other four procedures, i.e., $(\hat{R}_{OCBAbw} - \hat{R}_{others})/\hat{R}_{others}$, is represented in Table 5.11. Therefore, we can conclude that OCBAbw brings more benefits than the other four procedures do.

5.2.4 Summary

This section presents a procedure for selecting the best and worst designs simultaneously out of $\hat{k}$ designs with high efficiency. To efficiently select the best and worst designs, this section sets an objective function to maximize *Probability of Correct Selection* subject to the finite computing budget. Similar to [1, 2, 17], the procedure allocates replications in an asymptotic way to optimally approximate the true allocation sequence, which is a locally optimal procedure to achieve the goal.

To demonstrate the performance of the proposed algorithm, this section compares the proposed algorithm with four other different allocation procedures in seven distinct environments, including Equal, PTV, HalfT, and OCBAht. Equal and PTV serve as the benchmark of experiments. Although the second stage of HalfT shares the reserved information from the first stage, its efficiency is worse than OCBAbw's. As for heuristic OCBAht, the selection of the best and worst designs together is to take an average on the allocations of two situations in which the best or worst design is regarded as a target by directly applying OCBA, respectively. Numerical experimental results show that OCBAbw outperforms the others. The philosophy of selecting the best and worst simultaneously indicates tremendous potential. Furthermore, its applications to various fields [31–39] should be pursued.

5.3 Optimal Computing-Budget Allocation for Subset Ranking

Variants of OCBA are restricted in selecting the best design or optimal subset of designs or some specified designs like the best and worst introduced in Sections 5.1 and 5.2. However, a highly challenging issue, i.e., subset ranking, remains unaddressed. It goes beyond best design and optimal subset problems. This section develops a new OCBA-based approach to address the issue. Through necessary mathematical deduction, its theoretical foundation is laid. The numerical testing results show that, with proper parameters, it can indeed enhance the simulation efficiency and outperform other existing methods in terms of the probability of correct subset ranking and computational efficiency.

5.3.1 Background and Motivation

OCBA variants aim to select the best one or better ones out of all designs with various backgrounds. From another perspective, they separate the designs into two classes: one class contains the better designs (noticing that the class may contain only the best design) and the other one contains the worse designs. Formally, the set of designs is separated into two exclusive subsets, and any design in one subset is not worse than any one in the other. However, there are applications in practice where designs need to be separated into more classes. That is, the set needs to be separated into $\hat{n}$ exclusive subsets and in any two subsets, any design in one subset is not worse than any one in the other. The extreme case is that the number of subsets equals the number of designs: all designs need to be ranked accurately.

The best–worst scaling is a famous discrete choice model first described by Louviere [40] and requires three subsets. This model demands picking the item with the highest utility and the one with the lowest utility in a series of blocks containing many items [41], and is successfully used in many fields [41–44]. The question about how to guide potential customers to select and rank three subsets (one containing the best, one containing the worst, and the rest constitute the mediocre) correctly seems highly interesting. Such cases are encountered if the better designs are expected to have reward and the worse ones to have penalty or be discarded, while all the mediocre ones are retained. In many optimization processes, when a simulation-based setting is used in stochastic environments, we often need to reward better individuals while penalizing worse ones no matter what kinds of evolutionary computing models. As an Ant Colony Optimization (ACO) model, the Best–Worst Ant System (BWAS) strengthens the best solution and weakens the worst one while keeping the mediocre ones untouched [45]. In the Genetic Algorithm (GA) model, one of selection schemes is to push the best individual directly into the next generation without operations and eliminate the worst ones [23]. Similar concepts that elite particles replace the worse ones, while maintaining average ones, appear in Particle Swarm Optimization (PSO) [46–53].

In some situations, ranking every design is of benefit to understand each design's priority. It is an old problem and common in many fields. Ranking alternatives or options in the fuzzy set and system [54–59] is essential for preference ranking and fuzzy decision-making. Attention is also paid to ranking alternatives or options in linear programming [60], environmental appraisal [61], and information retrieval [62]. When these kinds of problems are set under stochastic environments, optimal budget allocation for each design is desired. A summary of real simulation-based situations that require to address subset ranking issues is shown in Table 5.12.

Motivated by many important web-search applications, subset ranking has already been a noticed problem. Cossock and Zhang study it [63, 64] and successfully use regression to rank subsets. Their work [65] presents another computationally more tractable approach to the problem based on proposed bounds of discounted cumulated gain. Handling such issues is also a key to feature extraction, construction, and selection problems [66, 67], which are essential pre-processing steps for problems on machine learning and pattern recognition. Once they are put into a simulation-based setting, one has to answer the question about how to do the desired budget allocation.

Table 5.12 Different numbers of classes in real situations.

The number of classes	Real situations
Two classes	Please see OCBAm [17]
Three classes	Best–worst scaling, global optimization, and so on
$\hat{k}$ classes where $\hat{k}$ is the number of designs	All designs need to be accurately ranked [54–62]

The objective is to find a simulation budget allocation policy maximizing the probability of correct subset ranking. The subset ranking issue is how to decompose Θ into $\hat{n}$ subsets such that $J_{j_1} \leq J_{j_2} \leq \cdots \leq J_{j_{\hat{n}}}$, $\forall j_i \in \dot{S}_i$. Therefore, *Correct Selection* is defined by

$$\text{CS} = \left\{ J_{j_1} \leq J_{j_2} \leq \cdots \leq J_{j_{\hat{n}}}, \ \forall j_i \in \dot{S}_i \right\}, \tag{5.48}$$

and to maximize *Probability of Correct Selection* is given as:

$$\max P\{\text{CS}\}$$
$$\text{s.t.} \sum_{i=1}^{\hat{k}} N_i = \bar{T}, \quad N_i \in \mathbb{Z}^+. \tag{5.49}$$

5.3.2 Approximate Optimal Simulation Budget Allocation

In the Bayesian framework, the probability of (5.48) is given as $P\{\text{CS}\} = P\left\{ \tilde{J}_{j_1} \leq \tilde{J}_{j_2} \leq \cdots \leq \tilde{J}_{j_{\hat{n}}}, \ \forall j_i \in \dot{S}_i \right\}$. $P\{\text{CS}\}$ is multiple integrations of different normal probability density functions, and the lower bound of one design integral is based on values of other designs. More accurately, the lower bound is the maximum value among values of all designs whose values are less than the current design. As a consequence, multiple integrals and max operations from the expansion prevent any easy calculation of its exact value. However,

$$P\left\{ \tilde{J}_{j_1} \leq \tilde{J}_{j_2} \leq \cdots \leq \tilde{J}_{j_{\hat{n}}}, \ \forall j_i \in \dot{S}_i \right\}$$
$$\geq P\left\{ \tilde{J}_{j_1} \leq \hat{c}_1 \leq \tilde{J}_{j_2} \leq \hat{c}_2 \leq \cdots \leq \hat{c}_{\hat{n}-1} \leq \tilde{J}_{j_{\hat{n}}}, \ \forall j_i \in \dot{S}_i \right\}$$
$$= \text{APCS},$$

where APCS is a lower bound of $P\{\text{CS}\}$ with given constants $\hat{c}_1, \hat{c}_2, \ldots,$ and $\hat{c}_{\hat{n}-1}$, called performance division parameters. Due to the difficulty to calculate $P\{\text{CS}\}$, the goal to maximize APCS instead of $P\{\text{CS}\}$ is reasonable, and thus (5.49) can be switched to

$$\max \text{APCS}$$
$$\text{s.t.} \sum_{i=1}^{\hat{k}} N_i = \bar{T}, \quad N_i \in \mathbb{Z}^+. \tag{5.50}$$

Although an analytic expression for the tightness of APCS is not possible so far due to the difficulty to compute $P\{\text{CS}\}$, its estimation can be extracted from numerical results to demonstrate the applicability of APCS.

In this part, a theorem of an asymptotically optimal solution to subset ranking is firstly introduced and then proved.

Theorem 5.3 *Given the total number of samples $\bar{T}$ to be allocated to $\hat{k}$ designs whose performances are depicted by random variables with means J_i and variances $\hat{\sigma}_i^2$, as $\bar{T} \to +\infty$, APCS is approximately maximized if $\forall \hat{h}_1, \hat{h}_2 \in \Theta$,*

$$\frac{\mathbb{N}_{\hat{h}_1}}{\mathbb{N}_{\hat{h}_2}} = \frac{\hat{\sigma}_{\hat{h}_1}^2 / \min\left(\left(\hat{c}_{\hat{g}_1} - \vec{J}_{\hat{h}_1}\right)^2, \left(\hat{c}_{\hat{g}_1 - 1} - \vec{J}_{\hat{h}_1}\right)^2\right)}{\hat{\sigma}_{\hat{h}_2}^2 / \min\left(\left(\hat{c}_{\hat{g}_2} - \vec{J}_{\hat{h}_2}\right)^2, \left(\hat{c}_{\hat{g}_2 - 1} - \vec{J}_{\hat{h}_2}\right)^2\right)},$$

where $\hat{c}_{\hat{g}_i}$ and $\hat{c}_{\hat{g}_i - 1}$ are the adjacent c values of $\hat{h}_i$ $i = 1, 2$.

$\tilde{J}_i$*'s are considered mutually independent. With $\hat{c}_0 = -\infty$ and $\hat{c}_{\hat{n}} = +\infty$, APCS is denoted as*

$$APCS = \prod_{i=1}^{\hat{n}} \prod_{j \in \dot{S}_i} P\left\{\hat{c}_{i-1} \le \tilde{J}_j \le \hat{c}_i\right\}. \tag{5.51}$$

DeGroot [68] has shown the fact that $\tilde{J}_j \sim \mathcal{N}\left(\vec{J}_j, \frac{\hat{\sigma}^2}{\mathbb{N}_j}\right)$, $\forall j \in \Theta$, which means that $\frac{\left(\tilde{J}_j - \vec{J}_j\right)}{\hat{\sigma}_j / \sqrt{\mathbb{N}_j}} \sim \mathcal{N}(0, 1)$, $\forall j \in \Theta$. So, for any $i \in \{1, 2, \dots, \hat{n}\}$ and for any $j \in \dot{S}_i$,

$$P\left\{\hat{c}_{i-1} \le \tilde{J}_j \le \hat{c}_i\right\} = \tilde{\Phi}\left(\frac{\hat{c}_i - \vec{J}_j}{\hat{\sigma}_j / \sqrt{\mathbb{N}_j}}\right) - \tilde{\Phi}\left(\frac{\hat{c}_{i-1} - \vec{J}_j}{\hat{\sigma}_j / \sqrt{\mathbb{N}_j}}\right). \tag{5.52}$$

By substituting (5.52) into (5.51) and introducing a Lagrangian multiplier for solving (5.50), we have

$$\bar{F} = \prod_{i=1}^{\hat{n}} \prod_{j \in \dot{S}_i} \left(\tilde{\Phi}\left(\frac{\hat{c}_i - \vec{J}_j}{\hat{\sigma}_j / \sqrt{\mathbb{N}_j}}\right) - \tilde{\Phi}\left(\frac{\hat{c}_{i-1} - \vec{J}_j}{\hat{\sigma}_j / \sqrt{\mathbb{N}_j}}\right)\right)$$
$$- \bar{\lambda}\left(\sum_{i=1}^{\hat{k}} \mathbb{N}_i - \bar{T}\right).$$

Furthermore, KKT conditions [69] of (5.50) can be stated as follows:

$$\frac{\partial \bar{F}}{\partial \mathbb{N}_{\hat{h}}} = \left(\tilde{\Phi}\left(\frac{\hat{c}_{\hat{g}} - \vec{J}_{\hat{h}}}{\hat{\sigma}_{\hat{h}} / \sqrt{\mathbb{N}_{\hat{h}}}}\right) - \tilde{\Phi}\left(\frac{\hat{c}_{\hat{g} - 1} - \vec{J}_{\hat{h}}}{\hat{\sigma}_{\hat{h}} / \sqrt{\mathbb{N}_{\hat{h}}}}\right)\right)'$$
$$\cdot \prod_{i=1}^{\hat{n}} \prod_{j \in \dot{S}_i \setminus \{\hat{h}\}} \left(\tilde{\Phi}\left(\frac{\hat{c}_i - \vec{J}_j}{\hat{\sigma}_j / \sqrt{\mathbb{N}_j}}\right) - \tilde{\Phi}\left(\frac{\hat{c}_{i-1} - \vec{J}_j}{\hat{\sigma}_j / \sqrt{\mathbb{N}_j}}\right)\right)$$
$$- \bar{\lambda} = 0, \quad \forall \hat{h} \in \dot{S}_{\hat{g}}, \quad \forall \hat{g} = 1, 2, \dots, \hat{n}. \tag{5.53}$$

To simplify the notation, we let $\tilde{\delta}\left(\mathbb{N}_j\right) = \tilde{\Phi}\left(\frac{\hat{c}_i-\bar{J}_j}{\hat{\sigma}_j/\sqrt{\mathbb{N}_j}}\right) - \tilde{\Phi}\left(\frac{\hat{c}_{i-1}-\bar{J}_j}{\hat{\sigma}_j/\sqrt{\mathbb{N}_j}}\right)$ where $j \in \dot{S}_i$.
Hence, (5.53) can be rewritten as

$$\frac{\partial \bar{F}}{\partial \mathbb{N}_{\hat{h}}} = \tilde{\delta}'\left(\mathbb{N}_{\hat{h}}\right) \cdot \prod_{i=1}^{\hat{n}} \prod_{j\in \dot{S}_i\backslash\{\hat{h}\}} \tilde{\delta}\left(\mathbb{N}_j\right) - \bar{\lambda} = 0. \tag{5.54}$$

Due to (5.54), for any $\hat{h}_1 \in \dot{S}_{\hat{g}_1}$, $\hat{h}_2 \in \dot{S}_{\hat{g}_2}$ where $\hat{g}_1, \hat{g}_2 \in \{1, 2, \ldots, \hat{n}\}$ and $\hat{h}_1 \neq \hat{h}_2$, we have

$$\tilde{\delta}'\left(\mathbb{N}_{\hat{h}_1}\right) \cdot \prod_{i=1}^{\hat{n}} \prod_{j\in \dot{S}_i\backslash\{\hat{h}_1\}} \tilde{\delta}\left(\mathbb{N}_j\right) = \tilde{\delta}'\left(\mathbb{N}_{\hat{h}_2}\right) \cdot \prod_{i=1}^{\hat{n}} \prod_{j\in \dot{S}_i\backslash\{\hat{h}_2\}} \tilde{\delta}\left(\mathbb{N}_j\right). \tag{5.55}$$

On one hand, when $\prod_{i=1}^{\hat{n}} \prod_{j\in \dot{S}_i\backslash\{\hat{h}_1,\hat{h}_2\}} \tilde{\delta}\left(\mathbb{N}_j\right) \neq 0$ holds, then the below relation can be easily obtained:

$$\tilde{\delta}'\left(\mathbb{N}_{\hat{h}_1}\right)\tilde{\delta}\left(\mathbb{N}_{\hat{h}_2}\right) = \tilde{\delta}\left(\mathbb{N}_{\hat{h}_1}\right)\tilde{\delta}'\left(\mathbb{N}_{\hat{h}_2}\right). \tag{5.56}$$

On the other hand, if $\prod_{i=1}^{\hat{n}} \prod_{j\in \dot{S}_i\backslash\{\hat{h}_1,\hat{h}_2\}} \tilde{\delta}\left(\mathbb{N}_j\right) = 0$ is true, (5.55) must hold regardless of the relation between $\mathbb{N}_{\hat{h}_1}$ and $\mathbb{N}_{\hat{h}_2}$ and the relation between $\mathbb{N}_{\hat{h}_1}$ and $\mathbb{N}_{\hat{h}_2}$ can be the same as (5.56).

Then, let $\mathbb{N}_{\hat{h}} = \hat{\alpha}_h T$ where $\hat{\alpha}_{\hat{h}}$ can be regarded as the budget proportion for design $\hat{h}$ within the total budget $\bar{T}$. Obviously, $\hat{\alpha} \in (0, 1)$. Notice that $\tilde{\delta}\left(\mathbb{N}_{\hat{h}}\right)$ is a non-negative, increasing and bounded function, and

$$\tilde{\delta}'\left(\mathbb{N}_{\hat{h}}\right) = \frac{\hat{c}_{\hat{g}} - \bar{J}_{\hat{h}}}{2\sqrt{2\pi}\hat{\sigma}_{\hat{h}}\sqrt{\hat{\alpha}_h T}} e^{-\frac{\left(\hat{c}_{\hat{g}}-\bar{J}_{\hat{h}}\right)^2}{2\hat{\sigma}_{\hat{h}}^2/\hat{\alpha}_{\hat{h}}}\bar{T}}$$
$$+ \frac{\bar{J}_{\hat{h}} - \hat{c}_{\hat{g}-1}}{2\sqrt{2\pi}\hat{\sigma}_{\hat{h}}\sqrt{\hat{\alpha}_h T}} e^{-\frac{\left(\hat{c}_{\hat{g}-1}-\bar{J}_{\hat{h}}\right)^2}{2\hat{\sigma}_{\hat{h}}^2/\hat{\alpha}_{\hat{h}}}\bar{T}}. \tag{5.57}$$

Now, according to (5.57), we convert (5.56) to

$$\frac{\hat{c}_{\hat{g}_1} - \bar{J}_{\hat{h}_1}}{2\sqrt{2\pi}\hat{\sigma}_{\hat{h}_1}\sqrt{\hat{\alpha}_{\hat{h}_1}}} \tilde{\delta}\left(\mathbb{N}_{\hat{h}_2}\right) \cdot e^{-\frac{\left(\hat{c}_{\hat{g}_1}-\bar{J}_{\hat{h}_1}\right)^2}{2\hat{\sigma}_{\hat{h}_1}^2/\hat{\alpha}_{\hat{h}_1}}T}$$

$$+ \frac{\bar{J}_{\hat{h}_1} - \hat{c}_{\hat{g}-1}}{2\sqrt{2\pi}\hat{\sigma}_{\hat{h}_1}\sqrt{\hat{\alpha}_{\hat{h}_1}}} \tilde{\delta}\left(\mathbb{N}_{\hat{h}_2}\right) \cdot e^{-\frac{\left(\hat{c}_{\hat{g}_1-1}-\bar{J}_{\hat{h}_1}\right)^2}{2\hat{\sigma}_{\hat{h}_1}^2/\hat{\alpha}_{\hat{h}_1}}\bar{T}}$$

$$= \frac{\hat{c}_{\hat{g}_2} - \bar{J}_{\hat{h}_2}}{2\sqrt{2\pi}\hat{\sigma}_{\hat{h}_2}\sqrt{\hat{\alpha}_{\hat{h}_2}}} \tilde{\delta}\left(\mathbb{N}_{\hat{h}_1}\right) \cdot e^{-\frac{\left(\hat{c}_{\hat{g}_2}-\bar{J}_{\hat{h}_2}\right)^2}{2\hat{\sigma}_{\hat{h}_2}^2/\hat{\alpha}_{\hat{h}_2}}\bar{T}}$$

$$+ \frac{\vec{J}_{\hat{h}_2} - \hat{c}_{\hat{g}_2 - 1}}{2\sqrt{2\pi}\hat{\sigma}_{\hat{h}_2}\sqrt{\hat{\alpha}_{\hat{h}_2}}} \tilde{\delta}\left(\mathbb{N}_{\hat{h}_1}\right) \cdot e^{-\frac{\left(\hat{c}_{\hat{g}_2 - 1} - \vec{J}_{\hat{h}_2}\right)^2}{2\hat{\sigma}_{\hat{h}_2}^2/\hat{\alpha}_{\hat{h}_2}}\vec{T}}.$$ (5.58)

∎

The equality may be too strict to reach. However, we may obtain the relation by replacing equality with infinitesimal of the same order. Thus, before obtaining the relation between $\mathbb{N}_{\hat{h}_1}$ and $\mathbb{N}_{\hat{h}_2}$, we need to establish the following lemma.

Lemma 5.2 If $\grave{A}(x)e^{-\grave{a}x} + \grave{B}(x)e^{-\grave{b}x} \sim \grave{C}(x)e^{-\grave{c}x} + \grave{D}(x)e^{-\grave{d}x}$, where $\grave{A}(x), \grave{B}(x),$ $\grave{C}(x),$ and $\grave{D}(x)$ are non-negative, increasing, and bounded functions, and $\grave{a}, \grave{b}, \grave{c}, \grave{d} > 0,$ then $\min\left(\grave{a}, \grave{b}\right) = \min\left(\grave{c}, \grave{d}\right).$

Proof: Let $\grave{l} = \min\left(\grave{a}, \grave{b}\right)$ and $\grave{r} = \min\left(\grave{c}, \grave{d}\right)$ and define $\grave{L}(x)$ and $\grave{R}(x)$ as

$$\grave{L}(x) = \begin{cases} \grave{A}(x), & \grave{l} = \grave{a} \\ \grave{B}(x), & \grave{l} = \grave{b} \\ \grave{A}(x) + \grave{B}(x), & \grave{l} = \grave{a} = \grave{b} \end{cases}$$

$$\text{and } \grave{R}(x) = \begin{cases} \grave{C}(x), & \grave{r} = \grave{c} \\ \grave{D}(x), & \grave{r} = \grave{d} \\ \grave{C}(x) + \grave{D}(x), & \grave{r} = \grave{c} = \grave{d} \end{cases}$$

respectively. Obviously, $\grave{L}(x)$ and $\grave{R}(x)$ are non-negative, increasing, and bounded functions as well. Hence,

$$\grave{L}(x)e^{-\grave{l}x} \sim \grave{A}(x)e^{-\grave{a}x} + \grave{B}(x)e^{-\grave{b}x}$$ (5.59)

$$\text{and } \grave{R}(x)e^{-\grave{r}x} \sim \grave{C}(x)e^{-\grave{c}x} + \grave{D}(x)e^{-\grave{d}x}.$$ (5.60)

From (5.59), (5.60), and properties of $\grave{L}(x)$ and $\grave{R}(x)$, we obtain that $\grave{l} = \grave{r} \Rightarrow$ $\min\left(\grave{a}, \grave{b}\right) = \min\left(\grave{c}, \grave{d}\right).$ ∎

By Lemma 5.1 and the fact that $N \propto \alpha$, relation between $\mathbb{N}_{\hat{h}_1}$ and $\mathbb{N}_{\hat{h}_2}$ in (5.58) is

$$\frac{1}{\hat{\sigma}_{\hat{h}_1}^2/\hat{\alpha}_{\hat{h}_1}} \min\left(\left(\hat{c}_{\hat{g}_1} - \vec{J}_{\hat{h}_1}\right)^2, \left(\hat{c}_{\hat{g}_1 - 1} - \vec{J}_{\hat{h}_1}\right)^2\right)$$

$$= \frac{1}{\hat{\sigma}_{\hat{h}_2}^2/\hat{\alpha}_{\hat{h}_2}} \min\left(\left(\hat{c}_{\hat{g}_2} - \vec{J}_{\hat{h}_2}\right)^2, \left(\hat{c}_{\hat{g}_2 - 1} - \vec{J}_{\hat{h}_2}\right)^2\right)$$

$$\Rightarrow \frac{\mathbb{N}_{\hat{h}_1}}{\mathbb{N}_{\hat{h}_2}} = \frac{\hat{\sigma}_{\hat{h}_1}^2 / \min \left(\left(\hat{c}_{\hat{g}_1} - \vec{J}_{\hat{h}_1} \right)^2, \left(\hat{c}_{\hat{g}_1 - 1} - \vec{J}_{\hat{h}_1} \right)^2 \right)}{\hat{\sigma}_{\hat{h}_2}^2 / \min \left(\left(\hat{c}_{\hat{g}_2} - \vec{J}_{\hat{h}_2} \right)^2, \left(\hat{c}_{\hat{g}_2 - 1} - \vec{J}_{\hat{h}_2} \right)^2 \right)}. \tag{5.61}$$

If a solution satisfies (5.61), then KKT conditions must hold. According to the KKT sufficient condition, this solution is a local optimum to (5.50). Therefore, we have proved the theorem at the beginning of this part.

Additionally, due to (5.51) and (5.52) and $\mathbb{N}_i = \hat{\alpha}_i \bar{T}$, APCS has a certain value given $t = \bar{T}_0$ and, moreover, approaches one when the number of allocation budgets gets larger. Therefore, $P\{CS\}$, which is always larger than APCS, can be guaranteed by the latter. ∎

Based on Theorem 5.3, similarly to the previous literature like [1, 17, 20], presents a sequential approach to rank subsets with a given computing budget. Initially, $\bar{n}_0$ simulation replications for each design are conducted. As simulation proceeds, the sample mean and variance of each design are computed to approximate $\vec{J}_i$ and $\hat{\sigma}_i^2$, respectively. According to this collected simulation output, an incremental computing budget, $\dot{\Delta}$, is allocated at each stage. This procedure continues until the total budget $\bar{T}$ is exhausted. It is realized in Algorithm OCBA$_{SR}$

OCBA$_{SR}$ Algorithm

Input: Number of simulation replications for each design $\bar{n}_0$, number of designs $\hat{k}$, set of designs Θ, incremental computing budget $\dot{\Delta}$, total number of simulation replications $\bar{T}$.

Output: The best design.

1: Perform $\bar{n}_0$ simulation replications for all designs;
$\bar{l} \leftarrow 0$;
$\mathbb{N}_{1,\bar{l}} = \mathbb{N}_{2,\bar{l}} = \cdots = \mathbb{N}_{\hat{k},\bar{l}} = \bar{n}_0$.

2: If $\sum_{i \in \Theta} \mathbb{N}_{i,\bar{l}} \geq \bar{T}$, stop.

3: Increase the computing budget by an extra $\dot{\Delta}$;
Use Theorem 5.3 to compute the new budget allocation $\mathbb{N}_{1,\bar{l}+1}, \mathbb{N}_{2,\bar{l}+1}, \ldots,$ and $\mathbb{N}_{\hat{k},\bar{l}+1}$.

4: Simulate design i for additional $\max\left(0, \mathbb{N}_{i,\bar{l}+1} - \mathbb{N}_{i,\bar{l}}\right)$ replications, $i \in \Theta$;
$\bar{l} \leftarrow \bar{l} + 1$.

5: Go to step 2.

END

Note that we can calculate $\hat{n} - 1$ performance division parameters simultaneously instead of calculating one but breaking one set into two subsets for $\hat{n} - 1$ times. This is because that the latter way needs $\hat{n} - 1$ parameters as well, and it cannot increase the tightness between $P\{CS\}$ and APCS. Moreover, it brings the difficulty to determine how many budgets should be allocated to each breaking operation. But the former one can auto-adjust the attention paid to each parameter (corresponding to each breaking operation) according to calculated sample values at every iteration.

In Algorithm $OCBA_{SR}$, $\bar{l}$ is the iteration count. As simulation evolves, each design may change its belonging subset from iteration to iteration. However, when $\bar{T} \to +\infty$, which causes $\bar{l} \to +\infty$, each design will be in its true subordinate subset.

As performance division parameters of APCS, $\hat{c}_i$'s, $i = 1, 2, \ldots, \hat{n} - 1$, will impact the quality of APCS, the approximation of $P\{CS\}$ and thus influence the results drastically, because APCS is a lower bound of $P\{CS\}$. Choosing proper $\hat{c}_i$'s to make APCS as large as possible is helpful to deliver a better lower bound and get a well-maximized $P\{CS\}$.

When $\hat{n} = 2$, our algorithm is reduced into OCBAm [17]. In this case, only $\hat{c}_1$ is to be determined. From the suggestion in [17] where only one c value needs to be determined, we extend and apply it to our algorithm. Defining $\breve{\sigma}_j = \hat{\sigma}_j / \sqrt{\mathbb{N}_j}$ similarly to [17],

$$\hat{c}_i = \frac{\breve{\sigma}_{j_{\hat{m}+1}} \vec{J}_{j_{\hat{m}}} + \breve{\sigma}_{j_{\hat{m}}} \vec{J}_{j_{\hat{m}+1}}}{\breve{\sigma}_{j_{\hat{m}}} + \breve{\sigma}_{j_{\hat{m}+1}}}, \quad \forall i \in \{1, 2, \ldots, \hat{n} - 1\}, \tag{5.62}$$

where $\vec{J}_{j_{\hat{m}}}$ is the maximum $\vec{J}_j$ smaller than $\hat{c}_i$ and $\vec{J}_{j_{\hat{m}+1}}$ is the minimum $\vec{J}_j$ larger than $\hat{c}_i$.

Equation (5.62) may not be effective when $\hat{n} > 2$. This is because, once a design gets a large number of simulation replications (the sample mean of this design is close to one of its adjacent $\hat{c}_i$'s), its decreasing $\breve{\sigma}$ leads to more and more replications for it. Thus by observing values, we can see that the distance between the design and its adjacent $\hat{c}_i$ becomes shorter and shorter. It is undesirable that the number of replications allocated to other designs is reduced. As a result, they can hardly be distinguished. With this consideration, we try to restrict such influence to have more accurate APCS.

Notice that we use $\hat{c}_i$ to separate the distance between $\vec{J}_{j_{\hat{m}}}$ and $\vec{J}_{j_{\hat{m}+1}}$ into two segments. We denote the shorter segment with length $\tilde{L}_{i,\min}$ and the longer one with length $\tilde{L}_{i,\max}$. Equation (5.62) probably makes the ratio of $\tilde{L}_{i,\min}$ to $\tilde{L}_{i,\max}$ too small for some i. Thus, one means is to "move" several $\hat{c}_i$'s and accordingly ratios of $\tilde{L}_{i,\min}$ to $\tilde{L}_{i,\max}$ for any i may become equal. We provide two values for equalization: the maximum value and minimum value among all ratios. Forcibly, we let

$$\frac{\tilde{L}_{i,\min}}{\tilde{L}_{i,\max}} = \min_{\hat{h}} \frac{\tilde{L}_{\hat{h},\min}}{\tilde{L}_{\hat{h},\max}}, \quad \forall i \in \{1, 2, \ldots, \hat{n} - 1\} \tag{5.63}$$

$$\text{or } \frac{\tilde{L}_{i,\min}}{\tilde{L}_{i,\max}} = \max_{\hat{h}} \frac{\tilde{L}_{\hat{h},\min}}{\tilde{L}_{\hat{h},\max}}, \quad \forall i \in \{1, 2, \ldots, \hat{n} - 1\}. \tag{5.64}$$

We provide a simple way to let the value of each $\hat{c}_i$ equal the midpoint of $\vec{J}_{j_{\hat{m}}}$ and $\vec{J}_{J_{\hat{m}+1}}$, where $\vec{J}_{j_{\hat{m}}}$ is the largest value among the ones whose values are less than $\hat{c}_i$ and $\vec{J}_{j_{\hat{m}+1}}$ is the smallest value among the ones whose values are larger than $\hat{c}_i$. This "naive" way can also let all ratios of $\tilde{L}_{i,\min}$ to $\tilde{L}_{i,\max}$ equal for all i. Mathematically,

$$\hat{c}_i = \frac{\vec{J}_{j_{\hat{m}}} + \vec{J}_{j_{\hat{m}+1}}}{2}, \quad \forall i \in \{1, 2, \ldots, \hat{n}-1\}, \tag{5.65}$$

where $\vec{J}_{j_{\hat{m}}}$ and $\vec{J}_{j_{\hat{m}+1}}$ have the same meanings as mentioned.

5.3.3 Simulation Studies

To test and compare with different allocation procedures, we conduct numerical experiments in eight distinct environments. Instead of $P\{CS\}$ and APCS, FCS (*Frequency of Correct Selection*) is used to show performances of the tested algorithms.

Equal allocation is a naive way to select the target. Samples are allocated equally to each design. Namely, each one has $\bar{T}/\hat{k}$ replications. According to [1, 17], it serves as a benchmark for comparison.

PTV procedure sequentially determines, at every iteration, the number of simulation replications for each design based on the newly updated sample variance. All designs are simulated for $\bar{n}_0$ replications at first, and then at every iteration, the ratios among all designs are calculated to best satisfy $\frac{N_1^{\bar{l}+1}}{\hat{s}_1^2} = \frac{N_2^{\bar{l}+1}}{\hat{s}_2^2} = \cdots = \frac{N_k^{\bar{l}+1}}{\hat{s}_k^2}$ where $\sum N_i^{\bar{l}+1} = \sum N_i^{\bar{l}} + \Delta$. Thus, design i is conducted for $\max\left(0, N_i^{\bar{l}+1} - N_i^{\bar{l}}\right)$ replications at the $(\bar{l}+1)$th iteration.

OCBA is a sequential procedure for selecting the best design. Its efficiency has been demonstrated in many fields, but it is not suitable to select and rank all the subsets.

According to the above discussion of $\hat{c}_i$, we try four versions of OCBA$_{SR}$: OCBA$_{SR}$ with (5.62) ("OCBA$_{SR}$m" for short), with (5.63) ("OCBA$_{SR}$min" for short), with (5.64) ("OCBA$_{SR}$max" for short), and with a midpoint (5.65) ("OCBA$_{SR}$mp" for short).

Except "Equal" (short for Equal Allocation), each procedure needs an initial budget $\bar{n}_0$ and an extra budget Δ in every iteration. We set $\bar{n}_0 = \Delta = 50$. Totally eight environments are taken as examples for illustrating the efficiency of the algorithm: Environments 5.5 and 5.6 about equal division of alternative designs (with equal and unequal variance, respectively); Environments 5.7 and 5.8 about unequal division (with equal and unequal variances, respectively); Environments 5.9 and 5.10 about larger-scale cases (larger number of subsets and designs, respectively); Environments 5.11 and 5.12 about extreme cases (with equal and unequal variances, respectively). Because OCBA$_{SR}$ will be reduced into OCBAm [17] if $\hat{n} = 2$, cases to rank only two subsets are excluded.

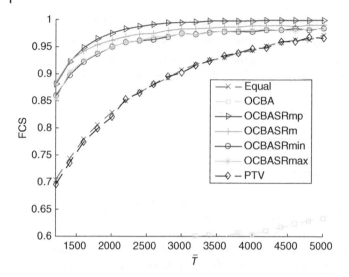

Figure 5.10 FCS versus $\bar{T}$ with different procedures in Environment 5.5.

Environment 5.5 (Equal Division and Equal Variance): In this case, $\hat{k} = 15$ and design i obeys $\mathcal{N}\left(i, 6^2\right)$. Notice that each design has the same variance. Designs are to be separated into three classes and $|\dot{S}_1| = |\dot{S}_2| = |\dot{S}_3| = 5$ (three subsets with the equal number of elements). From Fig. 5.10, it can be seen that all procedures obtain higher FCS as the available computing budget increases. Among them, four kinds of OCBA$_{SR}$ achieve the fastest speed to reach high FCS with a lower amount of replications than others. In this case, numeric results reveal that four OCBA$_{SR}$ procedures outperform the others.

Environment 5.6 (Equal Division and Unequal Variance): It is necessary to conduct experiments to test when the variances are not equal in each design. Fifteen alternative designs are with distribution $\mathcal{N}\left(i, i^2\right)$ for design i. Subsets are set as same as Environment 5.5. From Fig. 5.11, as available computing budget increases, all procedures obtain higher FCS. In this case, four kinds of OCBA$_{SR}$ procedures still are the fastest and outperform the others.

Environment 5.7 (Unequal Division and Equal Variance): $\hat{k} = 15$ in this case and design i obeys $\mathcal{N}\left(i, 6^2\right)$. Designs are expected to be separated into three classes: the first subset containing three better designs, the second containing five average one, and the third containing seven worse ones, i.e., $|\dot{S}_1| = 3$, $|\dot{S}_2| = 5$, and $|\dot{S}_3| = 7$. Each subset has a different number of designs from others. From Fig. 5.12, it can be seen that all procedures obtain higher FCS as the available computing budget increases. Among all methods, four kinds of

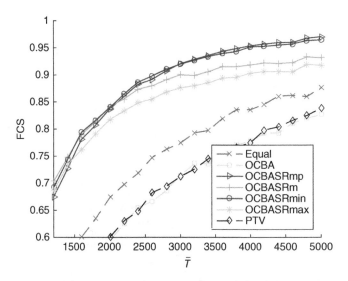

Figure 5.11 FCS versus $\bar{T}$ with different procedures in Environment 5.6.

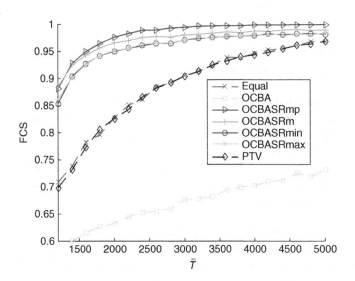

Figure 5.12 FCS versus $\bar{T}$ with different procedures in Environment 5.7.

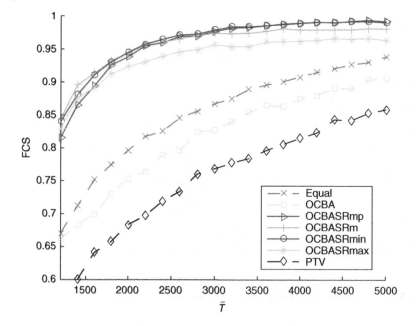

Figure 5.13 FCS versus $\bar{T}$ with different procedures in Environment 5.8.

$OCBA_{SR}$ procedures still achieve the fastest speed to reach high FCS with a lower amount of replications than others. In this case, numeric results reveal again that four $OCBA_{SR}$ procedures outperform other ones.

Environment 5.8 (Unequal Division and Unequal Variance): There are 15 alternative designs with distribution $\mathcal{N}\left(i, i^2\right)$ for design i. The numbers of subsets is expected to be set as the same as Environment 5.7. Similarly, from Fig. 5.13, all procedures obtain higher FCS as the available computing budget increases. Among them, four $OCBA_{SR}$ ones achieve the fastest speed to reach high FCS with a lower amount of replications than others. In this case, numeric results show that they outperform the other three procedures.

Environment 5.9 (Larger Number of Divisions): For $OCBA_{SR}$ that can be applied to any number of divisions, we show performances under the case with a larger number of division. There are 15 designs with distribution $\mathcal{N}\left(i, 6^2\right)$ for design i. Designs are required to be separated equally into five classes and $|\dot{S}_i| = 3$, $i = 1, 2, \ldots, 5$. From Fig. 5.14, except OCBA with low FCS that disappears from the figure, the other procedures obtain higher FCS as the available budget increases. Among them, the four $OCBA_{SR}$ ones achieve the fastest speed to reach high FCS with a lower amount of replications than others. Like the

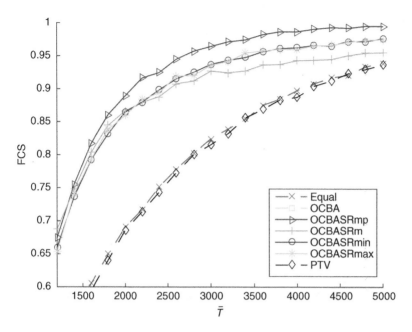

Figure 5.14 FCS versus $\bar{T}$ with different procedures in Environment 5.9.

above-mentioned cases, a similar conclusion about advantages of $OCBA_{SR}$ can be made for this case.

Environment 5.10 (Larger Number of Designs): It is also necessary to conduct experiments when there are a larger number of designs. In this case, $\hat{k} = 30$ and design i obeys $\mathcal{N}(i, 6^2)$. Designs are planned to be separated into three classes. Each one contains 10 elements, i.e., $|\dot{S}_1| = |\dot{S}_2| = |\dot{S}_3| = 10$. Except the disappeared OCBA, Fig. 5.15 similarly shows that procedures obtain higher FCS as the available budget increases. Four kinds of $OCBA_{SR}$ among all procedures achieve the fastest speed to reach high FCS with a lower amount of replications than others. In this case, experimental results reveal that this case still has similar conclusion like Environment 5.9.

Environment 5.11 (Extreme Case with Equal Variance): It is necessary to verify $OCBA_{SR}$ under extreme cases. In the first extreme case, 15 alternative designs are with distribution $\mathcal{N}(i, 6^2)$ for design i. All designs are expected to be ranked accurately. From Fig. 5.16, all procedures still obtain higher FCS as the budget increases. Among them, $OCBA_{SR}$mp achieves the fastest speed to reach high FCS with a lower amount of replications than others. $OCBA_{SR}$min and $OCBA_{SR}$max have a faster speed than Equal, PTV, and OCBA. Though the

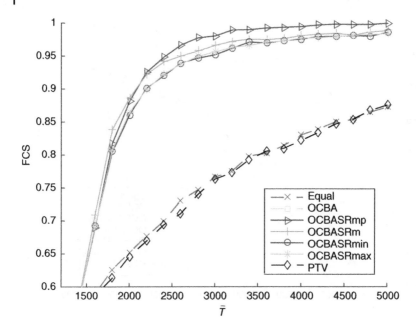

Figure 5.15 FCS versus $\bar{T}$ with different procedures in Environment 5.10.

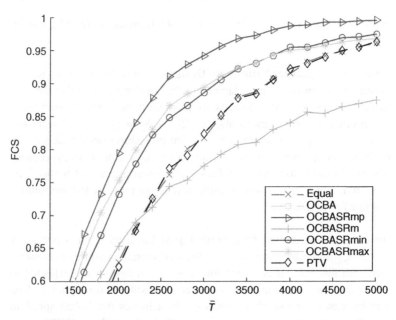

Figure 5.16 FCS versus $\bar{T}$ with different procedures in Environment 5.11.

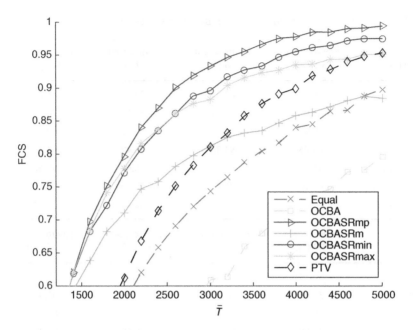

Figure 5.17 FCS versus $\bar{T}$ with different procedures in Environment 5.12.

superiority is not so significant as the previous cases, it represents a surprising result. This is because, in the extreme case, each design should be paid attention to, and Equal and PTV indeed pay attention to each one. However, the performance of $OCBA_{SR}m$ is not satisfactory. This is because some design is close to some $\hat{c}_i$ and allocated with the most budget. So, such allocation like $OCBA_{SR}m$ is not reasonable.

Environment 5.12 (Extreme Case with Unequal Variance): Instead of equal variance, it is necessary to verify performance of $OCBA_{SR}$ under the extreme case with unequal variance. Fifteen alternative designs are with distribution $\mathcal{N}\left(i, i^2\right)$ for design i. The goal is the same as Environment 5.11. From Fig. 5.17, all procedures obtain higher FCS as the budget increases. The performance of $OCBA_{SR}mp$ is still the best. $OCBA_{SR}min$ and $OCBA_{SR}max$ are faster than Equal, PTV, and OCBA. Though the superiority is not so significant as the previous environments, it is still surprising for similar reasons of Environment 5.11. $OCBA_{SR}m$ does not perform well yet. Distinct from Environment 5.11, the performance of PTV is close to $OCBA_{SR}min$ and $OCBA_{SR}max$ thanks to unequal variance.

Table 5.13 FCS of procedures when that of OCBA$_{SR}$mp reaches 95%.

Procedure	Environment 5.5 (%)	Environment 5.6 (%)	Environment 5.7 (%)	Environment 5.8 (%)
Equal	80.65	83.55	79.64	81.70
OCBA	57.02	77.71	62.64	76.37
PTV	79.92	77.50	80.62	69.72
OCBA$_{SR}$mp	**96.46**	**95.36**	**96.47**	**95.47**
OCBA$_{SR}$m	**95.39**	91.81	**95.67**	**95.60**
OCBA$_{SR}$min	93.70	**95.18**	94.20	**95.73**
OCBA$_{SR}$max	93.82	90.30	94.24	93.03

Procedure	Environment 5.9 (%)	Environment 5.10 (%)	Environment 5.11 (%)	Environment 5.12 (%)
Equal	80.23	73.03	85.04	78.69
OCBA	37.68	57.24	21.06	65.91
PTV	80.03	71.07	85.23	85.79
OCBA$_{SR}$mp	**95.70**	**96.66**	**95.68**	**95.55**
OCBA$_{SR}$m	91.20	94.98	79.25	83.21
OCBA$_{SR}$min	92.52	93.92	90.58	92.76
OCBA$_{SR}$max	92.02	93.79	91.08	91.60

The bold values mean that their FCS > 95.

Above environments illustrate the comparison results between four kinds of OCBA$_{SR}$ and three other ones. Table 5.13 summarizes all the results and shows that when OCBA$_{SR}$mp reaches FCS = 95%, other procedures have not even reached 86% in each environment. To conclude, OCBA$_{SR}$mp outperforms Equal, PTV, and OCBA in the above cases. Although in most cases three other variants of OCBA$_{SR}$ perform closely to OCBA$_{SR}$mp, in extreme cases, they do perform in a satisfactory way. This may be due to the fact that, though OCBA$_{SR}$mp is more naive than other kinds of OCBA$_{SR}$, it pays more balanced attention to designs than the others. Hence, its performance is more stable than the others. On the other hand, the performance gap between OCBA$_{SR}$, and some procedures like Equal and PTV get smaller as the density of division increases. This is because, as the density of division increases, more designs should receive "equal" attentions and priorities, and thus OCBA$_{SR}$ concerns on each design more equally and performs more similarly to Equal and PTV.

5.3.4 Summary

While many variants of OCBA can be used to efficiently select the best design or optimal subset, they may perform badly to rank more than two subsets. This section extends the original OCBA and OCBAm, and presents a OCBA-based procedure for subset ranking with a high efficiency. To efficiently select and rank subsets, we set an objective function to maximize *Probability of Correct Selection* subject to the fact that the sum of all simulation replications equals the given budget. The procedure allocates simulation replications in a way that it improves an asymptotic approximation to *Probability of Correct Selection*, and it is a locally optimal procedure to achieve the goal. This section then presents a sequential approach to solve this subset selection and ranking problem. To demonstrate its performance, this section compares four kinds of OCBA$_{SR}$ with three different allocation procedures in eight distinct environments. Numeric results reveal that, with proper parameters, OCBA$_{SR}$ can outperform the others. Our future work intends to apply it to complex optimization problems [39, 70–74].

5.4 Exercises

1   What is the definition of Ordinal Optimization? What is the purpose of optimal computing-budget allocation?

2   Analyze the similarities and differences between learning automata and optimal computing-budget allocation.

3   In a discrete-event system composed of $\hat{k}$ decisions, the goal is to correctly select the best and worst decisions from the decision set $\Theta = \{1, 2, \dots, \hat{k}\}$. Let J_i be the mean value of each element. Prove that for each decision $i \in \Theta$, the posterior estimate $\tilde{J}_i$ obtained by sampling follows an i.i.d. (independent and identically distributed) normal distribution.

4   In a case that selecting the best design and the worst one with different weights, let two weights $\bar{r}_1 > 0$ and $\bar{r}_2 > 0$ represent the importance of selecting the best and worst designs, respectively, $\bar{r}_1 + \bar{r}_2 = 2$. A total number of simulation samples $\bar{T}$ are to be allocated to $\hat{k}$ competing designs. The performances of these designs are depicted by random variables with means J_i and variances $\hat{\sigma}_i^2$. As $\bar{T} \to +\infty$, prove that $APCS_r$ is asymptotical.

5   Write the pseudo code of weighted OCBAbw algorithm.

6 Briefly describe the subset ranking problem, and write its objective optimization function.

7 Considering the application scenarios with different subset divisions, please list the instances where two, three, and $\hat{k}$ subsets are divided.

8 As performance division parameters of APCS, $\hat{c}_i$, $i = 1, 2, \ldots, \hat{n} - 1$, will impact the quality of APCS, the approximation of $P\{CS\}$ and thus influence the results drastically. Discuss the suggested values of $\hat{c}_i$ when $\hat{n} = 2$ and $\hat{n} > 2$, respectively.

9 Applying the allocation ratio formula (4.30), prove that the optimal computing-budget allocation for the subset sorting has asymptotic local optimality.

10 Write the pseudo code for subset sorting using the optimal computation allocation algorithm.

References

1 H. C. Chen, C. H. Chen, and E. Yücesan, "Computing efforts allocation for ordinal optimization and discrete event simulation," *IEEE Transactions on Automatic Control*, vol. 45, no. 5, pp. 960–964, 2000.

2 C. H. Chen, J. Lin, E. Yücesan, and S. E. Chick, "Simulation budget allocation for further enhancing the efficiency of ordinal optimization," *Discrete Event Dynamic Systems*, vol. 10, no. 3, pp. 251–270, 2000.

3 B. Hruz and M. Zhou, "Modeling and control of discrete event dynamic systems," Springer, 2007.

4 M. Zhou and K. Venkatesh, "Modeling, simulation and control of flexible manufacturing systems: A Petri net approach," World Scientific, 1999.

5 A. M. Law and W. D. Kelton, "Simulation modeling and analysis," New York: McGraw-Hill, 1991.

6 Y. C. Ho, R. S. Sreenivas, and P. Vakili, "Ordinal optimization of Discrete Event Dynamic Systems," *Discrete Event Dynamic Systems*, vol. 2, no. 1, pp. 61–88, 1992.

7 Y. C. Ho, "An ordinal optimization approach to optimal control problems," *Automatica*, vol. 35, no. 2, pp. 331–338, 1999.

8 N. T. Patsis, C. H. Chen, and M. E. Larson, "SIMD parallel discrete-event dynamic system simulation," *IEEE Transactions on Control Systems Technology*, vol. 5, no. 1, pp. 30–41, 1996.

9 C. G. Cassandras, L. Dai, and C. G. Panayiotou, "Ordinal optimization for deterministic and stochastic discrete resource allocation," in *Proceedings of the 36th IEEE Conference on Decision and Control*, volume 1, pp. 662–667, 1997.

10 W. B. Gong, Y. C. Ho, and W. Zhai, "Stochastic comparison algorithm for discrete optimization with estimation," *SIAM Journal on Optimization*, vol. 10, no. 2, pp. 384–404, 2000.

11 H. Liu, Q. Zhao, N. Huang, and X. Zhao, "Production line capacity planning concerning uncertain demands for a class of manufacturing systems with multiple products," *IEEE/CAA Journal of Automatica Sinica*, vol. 2, no. 2, pp. 217–225, 2015.

12 C. H. Chen, K. Donohue, E. Yücesan, and J. Lin, "Optimal computing budget allocation for Monte Carlo simulation with application to product design," *Simulation Modelling Practice and Theory*, vol. 11, no. 1, pp. 57–74, 2003.

13 C. H. Chen and D. He, "Intelligent simulation for alternatives comparison and application to air traffic management," *Journal of Systems Science and Systems Engineering*, vol. 14, no. 1, pp. 37–51, 2005.

14 V. J. Romero, D. V. Ayon, and C. H. Chen, "Demonstration of probabilistic ordinal optimization concepts to continuous-variable optimization under uncertainty," *Optimization and Engineering*, vol. 7, no. 3, pp. 343–365, 2006.

15 Q. S. Jia, "Efficient computing budget allocation for simulation-based policy improvement," *IEEE Transactions on Automation Science and Engineering*, vol. 9, no. 2, pp. 342–352, 2012.

16 M. C. Fu, J. Q. Hu, C. H. Chen, and X. P. Xiong, "Simulation allocation for determining the best design in the presence of correlated sampling," *INFORMS Journal on Computing*, vol. 19, pp. 101–111, 2007.

17 C. H. Chen, D. He, M. Fu, and L. H. Lee, "Efficient simulation budget allocation for selecting an optimal subset," *INFORMS Journal on Computing*, vol. 20, no. 4, pp. 579–595, 2008.

18 W. P. Wong, W. Jaruphongsa, and L. H. Lee, "Budget allocation for effective data collection in predicting an accurate DEA efficiency score," *IEEE Transactions on Automatic Control*, vol. 56, no. 6, pp. 1235–1246, 2011.

19 L. H. Lee, N. A. Pujowidianto, L. W. Li, C.-H. Chen, and C. M. Yap, "Approximate simulation budget allocation for selecting the best design in the presence of stochastic constraints," *IEEE Transactions on Automatic Control*, vol. 57, no. 11, pp. 2940–2945, 2012.

20 Q. S. Jia, "Efficient computing budget allocation for simulation-based optimization with stochastic simulation time," *IEEE Transactions on Automatic Control*, vol. 58, no. 2, pp. 539–544, 2013.

21 Y. Peng, C. H. Chen, M. C. Fu, and J. Q. Hu, "Efficient simulation resource sharing and allocation for selecting the best," *IEEE Transactions on Automatic Control*, vol. 58, no. 4, pp. 1017–1023, 2013.

22 W. G Lou, "Study on layout planning of municipal sewage treatment plants by using osculating method," *Shanghai Environmental Sciences*, vol. 21, no. 6, pp. 327–329, 2002.

23 D. E. Goldberg and K. Deb, "A comparative analysis of selection schemes used in genetic algorithms," *Urbana*, vol. 51, pp. 61801–2996, 1991.

24 C. H. Chen, "A lower bound for the correct subset-selection probability and its application to discrete event system simulations," *IEEE Transactions on Automatic Control*, vol. 41, no. 8, pp. 1227–1231, 1996.

25 S. E. Chick, "Bayesian analysis for simulation input and output," in *Proceedings of the 1997 Winter Simulation Conference*, pp. 253–260, 1997.

26 S. R. Jaeger, A. S. Jørgensen, M. D. Aaslyng, and W. L. P. Bredie, "Best–worst scaling: An introduction and initial comparison with monadic rating for preference elicitation with food products," *Food Quality and Preference*, vol. 19, no. 6, pp. 579–588, 2008.

27 M. Mesterton-Gibbons "On the evolution of pure winner and loser effects: A game-theoretic model," *Bulletin of Mathematical Biology*, vol. 61, no. 6, pp. 1151–1186, 1999.

28 K. Hock and R. Huber, "Models of winner and loser effects: A cost-benefit analysis," *Behaviour*, vol. 146, no. 3, pp. 69–87, 2009.

29 G. X. Wei, "The comparison of innovation incentives between winner-up and loser-out tournaments," *Commercial Research*, 2006. URL: https://api .semanticscholar.org/CorpusID:167253348.

30 P. Zhang, S. Chen, and C. Zheng, "Practice of the first priority and the last reject system in annual comprehensive test appraisement of employees," *Chinese Nursing Research*, vol. 86, no. 5, pp. 1614–1618, 2007.

31 F.-Y. Wang, J. J. Zhang, X. Zheng, X. Wang, Y. Yuan, X. Dai, J. Zhang, and L. Yang, "Where does AlphaGo go: From church-turing thesis to AlphaGo thesis and beyond," *IEEE/CAA Journal of Automatica Sinica*, vol. 3, no. 2, pp. 113–120, 2016.

32 B. Liu, J. Li, C. Chen, W. Tan, Q. Chen, and M. C. Zhou, "Efficient motif discovery for large-scale time series in healthcare," *IEEE Transactions on Industrial Informatics*, vol. 11, no. 3, pp. 583–590, 2015.

33 P. Wu, A. Che, F. Chu, and M. C. Zhou, "An improved exact SA-constraint and cut-and-solve combined method for biobjective robust lane reservation," *IEEE Transactions on Intelligent Transportation Systems*, vol. 16, no. 3, pp. 1479–1492, 2015.

34 J. Li, M. C. Zhou, Q. Sun, X. Dai, and X. Yu, "Colored traveling salesman problem," *IEEE Transactions on Cybernetics*, vol. 45, no. 11, pp. 2390–2401, 2015.

35 Q. Kang, J. Wang, M. C. Zhou, and A. C. Ammari, "Centralized charging scheduling for electric vehicles with battery-swapping on spot pricing via heuristic algorithms," *IEEE Transactions on Intelligent Transportation Systems*, vol. 17, no. 3, pp. 659–669, 2016.

36 Q. Kang, S. Liu, M. Zhou, and S. Li, "A weight-incorporated similarity-based clustering ensemble method based on swarm intelligence," *Knowledge-Based Systems*, vol. 104, pp. 156–164, 2016.

37 Z. Ding, Y. Zhou, and M. C. Zhou, "Modeling self-adaptive software systems with learning Petri nets," *IEEE Transactions on Systems, Man, and Cybernetics: Systems*, vol. 46, no. 4, pp. 483–498, Apr 2016.

38 W. Han, J. Xu, M. C. Zhou, G. Tian, P. Wang, X. Shen, and S.-H. E. Hou, "Cuckoo-search and particle-filter-based inversing approach to estimating defects via magnetic flux leakage signals," *IEEE Transactions on Magnetics*, vol. 52, no. 4, Apr 2016, doi: 10.1109/TMAG.2015.2498119.

39 L. Dong, B. Shi, G. Tian, Y. B. Li, B. Wang, and M. C. Zhou, "An accurate de novo algorithm for glycan topology determination from mass spectra," *IEEE/ACM Transactions on Computational Biology and Bioinformatics*, vol. 12, no. 3, pp. 568–578, May-June 2015.

40 J. J. Louviere, "Best-Worst Scaling: A Model for the Largest Difference Judgments," Working Paper, University of Alberta, 1991.

41 T. N. Flynn, J. J. Louviere, T. J. Peters, and J. Coast, "Best–worst scaling: What it can do for health care research and how to do it," *Journal of Health Economics*, vol. 26, no. 1, pp. 171–189, 2007.

42 P. Auger, T. M. Devinney, and J. J. Louviere, "Using best-worst scaling methodology to investigate consumer ethical beliefs across countries," *Journal of Business Ethics*, vol. 70, no. 3, pp. 299–326, 2007.

43 J. A. Lee, G. N. Soutar, and J. Louviere, "Measuring values using best-worst scaling: The LOV example," *Psychology & Marketing*, vol. 24, no. 12, pp. 1043–1058, 2007.

44 T. N. Flynn, J. J. Louviere, T. J. Peters, and J. Coast, "Estimating preferences for a dermatology consultation using Best-Worst Scaling: Comparison of various methods of analysis," *BMC Medical Research Methodology*, vol. 8, no. 1, p. 76, 2008.

45 O. Cordon, I. F. de Viana, F. Herrera, and L. Moreno, "A new ACO model integrating evolutionary computation concepts: The best-worst Ant System," University Librede Bruxelles, Brussels, Belgium, pp. 22–29, 2000. URL: https://api.semanticscholar.org/CorpusID:16076129.

46 Y. H. Shi and R. C. Eberhart, "A modified particle swarm optimizer," in *Proceedings of the IEEE International Conference on Evolutionary Computation*, pp. 69–73, 1998.

47 P. C. Fourie and A. A. Groenwold, "The particle swarm optimization algorithm in size and shape optimization," *Structural and Multidisciplinary Optimization*, vol. 23, no. 4, pp. 259–267, 2002.

48 S. Zhang, P. Chen, L. H. Lee, C. E. Peng, and C.-H. Chen, "Simulation optimization using the particle swarm optimization with optimal computing budget allocation," in *Proceedings of the Winter Simulation Conference (WSC)*, 2011.

49 J. Li, J. Zhang, C. Jiang, and M. Zhou, "Composite particle swarm optimizer with historical memory for function optimization," *IEEE Transactions on Cybernetics*, vol. 45, no. 10, pp. 2350–2363, 2015.

50 J. Zhang, X. Zhu, Y. Wang, and M. Zhou, "Dual-environmental particle swarm optimizer in noisy and noise-free environments," *IEEE Transactions on Cybernetics*, vol. 49, no. 6, pp. 2011–2021, 2019.

51 J. Bi, H. Yuan, S. Duanmu, M. C. Zhou, and A. Abusorrah, "Energy-optimized partial computation offloading in mobile-edge computing with genetic simulated-annealing-based particle swarm optimization," *IEEE Internet of Things Journal*, vol. 8, no. 5, pp. 3774–3785, 2021.

52 H. Yuan and M. Zhou, "Profit-maximized collaborative computation offloading and resource allocation in distributed cloud and edge computing systems," *IEEE Transactions on Automation Science and Engineering*, vol. 18, no. 3, pp. 1277–1287, 2021.

53 H. Yuan, J. Bi, W. Tan, M. Zhou, B. H. Li, and J. Li, "TTSA: An effective scheduling approach for delay bounded tasks in hybrid clouds," *IEEE Transactions on Cybernetics*, vol. 47, no. 11, pp. 3658–3668, 2017.

54 J. J. Buckley, "Ranking alternatives using fuzzy numbers," *Fuzzy Sets and Systems*, vol. 15, no. 1, pp. 21–31, 1985.

55 Z. P. Fan, J. Ma, and Q. Zhang, "An approach to multiple attribute decision making based on fuzzy preference information on alternatives," *Fuzzy Sets and Systems*, vol. 131, no. 1, pp. 101–106, 2002.

56 Y. M. Wang, J. B. Yang, and D. L. Xu, "A two-stage logarithmic goal programming method for generating weights from interval comparison matrices," *Fuzzy Sets and Systems*, vol. 152, no. 3, pp. 475–498, 2005.

57 F. Torfi, R. Z. Farahani, and S. Rezapour, "Fuzzy AHP to determine the relative weights of evaluation criteria and Fuzzy TOPSIS to rank the alternatives," *Applied Soft Computing*, vol. 10, no. 2, pp. 520–528, 2010.

58 J. Guo, M. Zhou, Z. Li, and H. Xie, "Green design assessment of electromechanical products based on group weighted-AHP," *Enterprise Information Systems*, vol. 9, no. 8, pp. 878–899, 2015.

59 G. Tian, H. Zhang, M. Zhou, and Z. Li, "AHP, gray correlation, and TOPSIS combined approach to green performance evaluation of design alternatives," *IEEE Transactions on Systems, Man, and Cybernetics: Systems*, vol. 48, no. 7, pp. 1093–1105, 2018.

60 B. Malakooti, "Ranking multiple criteria alternatives with half-space, convex, and non-convex dominating cones: Quasi-concave and quasi-convex multiple attribute utility functions," *Computers & Operations Research*, vol. 16, no. 2, pp. 117–127, 1989.

61 M. G. Rogers and M. P. Bruen, "Using ELECTRE to rank options within an environmental appraisal-two case studies," *Civil Engineering Systems*, vol. 13, no. 3, pp. 203–221, 1996.

62 T. Y. Liu, "Learning to rank for information retrieval," *Foundations and Trends in Information Retrieval*, vol. 3, no. 3, pp. 225–331, 2009.

63 D. Cossock and T. Zhang, "Subset ranking using regression," in Learning Theory, pp. 605–619, 2006.

64 S. Greco, V. Mousseau, and R. Słowiński, "Ordinal regression revisited: Multiple criteria ranking using a set of additive value functions," *European Journal of Operational Research*, vol. 191, no. 2, pp. 416–436, 2008.

65 D. Cossock and T. Zhang, "Statistical analysis of Bayes optimal subset ranking," *IEEE Transactions on Information Theory*, vol. 54, no. 11, pp. 5140–5154, 2008.

66 J. Yang and V. Honavar, "Feature subset selection using a genetic algorithm," in Feature Extraction, Construction and Selection, pp. 117–136, 1998.

67 N. Kourosh and M. Zhang, "Genetic programming for feature subset ranking in binary classification problems," Genetic Programming, pp. 121–132, 2009.

68 M. H. DeGroot, "Optimal statistical decisions," McGraw-Hill, Inc., 1970.

69 R. C. Walker, "Introduction to mathematical programming," Prentice Hall, 1999.

70 A. Che, P. Wu, F. Chu, and M. C. Zhou, "Improved quantum-inspired evolutionary algorithm for large-size lane reservation," *IEEE Transactions on Systems, Man, and Cybernetics: Systems*, vol. 45, no.12, pp. 1535–1548, Dec 2015.

71 Z. J. Ding, J. J. Liu, Y. Q. Sun, C J. Jiang, and M. C. Zhou, "A transaction and QoS-aware service selection approach based on genetic algorithm," *IEEE Transactions on Systems, Man, and Cybernetics: Systems*, vol. 45, no. 7, pp. 1035–1046, July 2015.

72 X. Zuo, C. Chen, W. Tan, and M. C. Zhou, "Vehicle scheduling of urban bus line via an improved multi-objective genetic algorithm," *IEEE Transactions on Intelligent Transportation Systems*, vol. 16, no. 2, pp. 1030–1041, Apr 2015.

73 Q. Deng, Q. Kang, L. Zhang, M. Zhou, and J. An, "Objective space-based population generation to accelerate evolutionary algorithms for large-scale many-objective optimization," *IEEE Transactions on Evolutionary Computation*, vol. 27, no. 2, pp. 326–340, April 2023

74 Z. Wang, S. Gao, M. Zhou, S. Sato, J. Cheng, and J. Wang, "Information-theory-based nondominated sorting ant colony optimization for multiobjective feature selection in classification," *IEEE Transactions on Cybernetics*, vol. 53, no. 8, pp. 5276–5289, Aug. 2023

6

Incorporation of Ordinal Optimization into Learning Automata

In practical applications, a learning automaton acts as an adaptive controller in modeling a process as well as an appropriate control signal generator. Systems built with learning automata (LAs) have been successfully employed in many difficult learning situations. The existing LAs employ heuristics to update their action probability vectors and then use the vectors for ordinal optimization and determining the computing budget size. The action probability vector of learning automata plays two roles: (i) Deciding when it converges, i.e., total computing budget it has used; and (ii) Allocating computing budget among actions to identify the optimal one. These two intertwined roles lead to a problem: The computing budget mostly goes to the currently estimated optimal action due to its high action probability regardless whether such budget allocation can help identify the true optimal one or not. This chapter introduces a new class of LAs. They no longer use their action probability vector for computing budget allocation. Instead they use such vector only to determine if they converge but employ Optimal Computing Budget Allocation (OCBA) introduced in Chapter 5 to accomplish the allocation of computing budget in a way that maximizes the probability of identifying the true optimal action. The proof of its ε-optimality is presented. Simulation results verify its significant advantages over the existing best algorithms. Compared with the state-of-the-art methods in five popular environments, can the introduced LAs speed up the learning efficiency significantly? This chapter answers this important question.

6.1 Background and Motivation

The action probability vector is the soul of LA, and thus its update rule is critical to LA. Among the LA families, the Estimator Algorithms are proved to be fast, and Pursuit Algorithms [1–5] and Last-position Elimination Algorithm [6] are among the top ones. The discretized pursuit scheme DP_{RI} [1, 3], introduced in

Learning Automata and Their Applications to Intelligent Systems, First Edition.
JunQi Zhang and MengChu Zhou.
© 2024 The Institute of Electrical and Electronics Engineers, Inc. Published 2024 by John Wiley & Sons, Inc.

Chapter 2, is introduced to speed up the continuous one [2]. Another famous one is the Discretized Generalized Pursuit Algorithm (DGPA) [4], which pursues the actions that have higher reward estimates than the current chosen action instead of a single estimated optimal one. Based on the pursuit scheme and a stochastic estimator (SE), SE_{RI} is introduced to further speed up LA convergence by utilizing a Reward–Inaction paradigm in a stationary random environment [5]. Reverse to the above Pursuit Algorithms, Last-position Elimination-based Learning Automata (LELA) [6], as introduced in Chapter 3, penalizes the action graded the last in terms of the estimated performance by decreasing its action probability and eliminates it when its action probability is decreased to zero. It outperforms the Pursuit Algorithms [6].

In [7], a statistical measure, i.e., the lower bound approximation of the confidence probability, is combined with LA in a heuristic and hybrid way resulting in learning automata with Confidence Probability (LACP). It is then extended from single-teacher environment to multi-teacher one and applied to a deterioration production system with preventive maintenance in [8].

Distinct from the combination scheme in LACP, this chapter presents a separation scheme inspired from the need to solve the problem in traditional LA: The computing budget, or more often sampling budget in the LA domain, mostly goes to the currently estimated optimal action due to its high action probability regardless whether such budget allocation can help identify the true optimal action or not. The action probability vector in LA has two roles: (i) Determining if LA converges, i.e., the size of sampling budget; and (ii) Allocating sampling budget among actions to identify the optimal one, where only ordinal optimization is required. Ordinal optimization [9] has emerged as an efficient technique for simulation optimization. Its underlying philosophy is to obtain a good estimate of the optimal action or design while the accuracy of an estimate can be relatively low [9, 10]. Ordinal comparison is much faster for selecting a high-quality action or design in many practical situations if our goal is to find the best one rather than an accurate estimate of the best performance value.

This chapter introduces a new class of LAs named LA_{OCBA}. Its action probability vector only determines the total sampling budget in an adaptive way. OCBA [11, 12] is employed to accomplish the ordinal optimization. It allocates the sampling budget to actions in a way that maximizes the probability of selecting the optimal action correctly. Thus, the action probability vector and OCBA are used to decide the total sampling budget and perform ordinal optimization, respectively.

OCBA is introduced in the field of simulation optimization to enhance the efficiency of ordinal optimization for discrete-event simulation designs in [11]. An improved version is introduced in [12] by replacing the objective function with a better approximation that can be solved analytically. Furthermore, in this improved version, Chernoff's bounds are not used in the derivation and fewer assumptions are imposed. Thus, this improved version is used in [12] and called as OCBA. Interested readers can refer to an excellent book [13] which extensively presents its developments and applications in large-scale stochastic simulation and optimization.

In OCBA, the actions or designs are defined as those solutions whose performances are polluted by noise and demand for re-evaluation to obtain a more accurate function value. OCBA first evaluates the means and variances of the action performances via allocating the same amount of initial samples $\hat{B}_0$ to each action. Next, it allocates an additional budget $\hat{B}_\Delta$ sequentially based on the means and variances, both of which are updated once a response is received. Finally, the approximate probability of correct selection is asymptotically maximized when the budget is used up according to the following equations:

$$\frac{\mathbb{N}_i(t)}{\mathbb{N}_j(t)} = \left(\frac{\hat{\sigma}_i(t)/(\hat{\delta}_{i,m}(t))}{\hat{\sigma}_j(t)/(\hat{\delta}_{j,m}(t))}\right)^2, \quad \forall i,j \in \mathbb{Z}_r \quad \text{and} \quad i \neq j \neq m, \tag{6.1}$$

$$\mathbb{N}_m(t) = \hat{\sigma}_m(t)\sqrt{\sum_{i=1,i\neq m}^{r} \frac{\mathbb{N}_i^2(t)}{(\hat{\sigma}_i(t))^2}}, \tag{6.2}$$

where $\mathbb{N}_i(t)$ refers to the total number of evaluations given to action i at allocation time t, $\hat{d}_i(t)$ and $\hat{\sigma}_i(t)$ store its sample mean and variance, respectively, m stands for the best action with the lowest mean, and $\hat{\delta}_{i,m}(t) = \hat{d}_i(t) - \hat{d}_m(t)$.

Equations (6.1) and (6.2) cannot be directly used to allocate the total budget to actions. Let $\bar{\varsigma} = \sum \mathbb{N}_i(t) / \sum \frac{\hat{\sigma}_i^2(t)}{\hat{\delta}_{i,m}^2(t)}, i \in 1, 2, \ldots, r, i \neq m$. According to (6.1) and (6.2), we have

$$\mathbb{N}_i(t) = \frac{\hat{\sigma}_i^2(t)}{\hat{\delta}_{i,m}^2(t)}\bar{\varsigma}, \quad i \in 1, 2, \ldots, r \quad \text{and} \quad i \neq m. \tag{6.3}$$

$$\mathbb{N}_m(t) = \hat{\sigma}_m(t) \sum_{i=1,i\neq m}^{r} \frac{\hat{\sigma}_i(t)}{\hat{\delta}_{i,m}^2(t)}\bar{\varsigma}. \tag{6.4}$$

Then, we have

$$\mathbb{N}_m(t) = \frac{\mathbb{N}(t)\hat{\sigma}_m(t)\sum_{i=1,i\neq m}^{r}\frac{\hat{\sigma}_i(t)}{\hat{\delta}_{i,m}^2(t)}}{\hat{\sigma}_m(t)\sum_{i=1,i\neq m}^{r}\frac{\hat{\sigma}_i(t)}{\hat{\delta}_{i,m}^2(t)} + \sum_{i=1,i\neq m}^{r}\frac{\hat{\sigma}_i^2(t)}{\hat{\delta}_{i,m}^2(t)}}$$

$$= \mathbb{N}(t) \sum_{i=1,i\neq m}^{r} \frac{\hat{\sigma}_m(t)}{\hat{\sigma}_m(t) + \hat{\sigma}_i(t)}, \tag{6.5}$$

where $\mathbb{N}(t) = \sum_{i=1}^{r} \mathbb{N}_i(t)$ is the total budget.

For $i \neq m$, we have

$$
\mathbb{N}_i(t) = \frac{\mathbb{N}(t)\dfrac{\hat{\sigma}_i^2(t)}{\hat{\delta}_{i,m}^2(t)}}{\hat{\sigma}_m(t)\sum_{j=1,j\neq m}^{r}\dfrac{\hat{\sigma}_j(t)}{\hat{\delta}_{j,m}^2(t)} + \displaystyle\sum_{j=1,j\neq m}^{r}\dfrac{\hat{\sigma}_j^2(t)}{\hat{\delta}_{j,m}^2(t)}}. \tag{6.6}
$$

Hence, according to (6.5) and (6.6), the total budget can be allocated to each action. The source code and demonstrations of OCBA can be found in [14]. The main purpose of OCBA is to allocate limited computational budget among the actions at time t in a way that maximizes the probability of correctly selecting the optimal action.

6.2 Learning Automata with Optimal Computing Budget Allocation

This section introduces LA$_{OCBA}$ which uses OCBA to replace the role of an LA's action probability vector in allocating sampling budget to all actions. In order to show the differences between the introduced LA$_{OCBA}$ and the classic LA, a learning model [15, 16] is used to model and interpret them as shown in Fig. 6.1. It can be divided into four conceptual components: Sensor, State Critic, Learning Rules, and Actuator.

(1) *Sensors*: They provide the response from the environment to the chosen action.
(2) *State Critic*: It counts and tells the Learning Rules how well the actions are doing with respect to a fixed performance standard and determines when the algorithm terminates.
(3) *Learning Rules*: They are responsible for making improvements, which use feedback from the State Critic on how LA is doing and determine how the Actuator should be modified to do better in the future. If Learning Rules are willing to explore a little, and do some perhaps suboptimal actions in the short run, LA might discover much better actions for the long run.
(4) *Actuator*: It is responsible for suggesting actions that lead to new and informative experience. It chooses the next action according to the action probability vector and tries the chosen action in the environment.

As shown in Fig. 6.1, the action probability vector P in classical LA performs two functions in both State Critic and Learning Rules: One determines when the algorithm should be terminated in State Critic. The other determines the next action to be tried in Learning Rules. Both functions rely on P, thereby leading to a problem: The sampling budget mostly goes to the currently estimated optimal action due

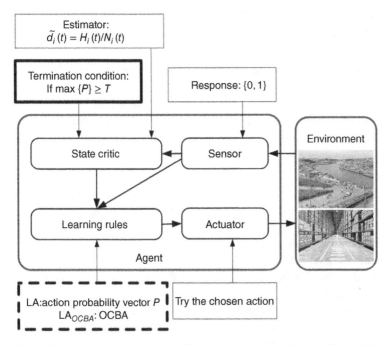

Estimator:
$$\tilde{d}_i(t) = H_i(t)/N_i(t)$$

Termination condition:
If max $\{P\} \geq T$

Response: $\{0, 1\}$

Environment

State critic

Sensor

Learning rules

Actuator

Agent

LA:action probability vector P Try the chosen action
LA$_{OCBA}$: OCBA

Figure 6.1 LA and LA$_{OCBA}$ illustrated in the framework of the learning model [15, 16]. Source: Petinovs/Adobe Stock Photos.

to its high action probability regardless whether such budget allocation can help identify the true optimal one or not.

Example 6.1 Let $D = \{0.75, 0.8, 0.6\}$ in dictating the true optimal action is α_2. $\tilde{D} = \{0.75, 0.7, 0.6\}$. $P = \{0.9, 0.05, 0.05\}$. Most samplings go to α_1 because $p_1 = 0.9$, which results in that the true optimal action α_2, whose reward probability is 0.8 but its estimated probability is only 0.7, is hard to be identified. This example shows that the allocation of sampling budget in traditional LA does not maximize the probability of identifying the true optimal action.

In the previous LA study, many heuristic schemes are introduced to improve the efficiency of identifying the true optimal action. For OCBA, the total budget is usually predetermined. In the introduced LA$_{OCBA}$, OCBA provides LA with a theoretical method for the allocation of sampling budget to identify the optimal action, while LA provides OCBA an adaptive way to determine the size of sampling budget. Introduction of OCBA into LA can thus advance LA in terms of its learning efficiency and accuracy improvement and OCBA in terms of its adaptability to the environment. The above ideas are realized in LA$_{OCBA}$ Algorithm.

LA$_{OCBA}$ Algorithm

Parameters

m Index of the maximal component of the reward estimate vector $\tilde{D}(t)$, $m = \text{argmax}_{i \in \mathbb{Z}_r} \{\tilde{d}_i(t)\}$;

$\mathbb{H}_i(t)$ Number of times the ith action has been rewarded up to time t, $i \in \mathbb{Z}_r$;

$\mathbb{N}_i(t)$ Number of times the ith action has been chosen up to time t, $\forall i \in \mathbb{Z}_r$;

n Resolution parameter;

$\Delta = \frac{1}{m}$ Step size.

Method

Input: Number of allowable actions r, resolution parameter n, action set A, environmental response set B, amount of initial samples $\hat{B}_0$, convergence threshold $\hat{T}$.

Output: Estimated optimal action α_m.

Initialize $t = 1$. $p_i(t) = 1/r, \forall i \in \mathbb{Z}_r$.

Initialize $\tilde{d}_i(t)$ by choosing each action for $\hat{B}_0$ times. $\mathbb{N}_1(t) = \mathbb{N}_2(t) = \cdots = \mathbb{N}_r(t) = \hat{B}_0$; $t = r\hat{B}_0$; $\hat{B}_\Delta(t) = \hat{B}_0$.

1: Increase the budget (i.e., the number of additional sampling for all actions) by $\hat{B}_\Delta(t)$ and let $\mathbb{N}(t) = \sum_{i=1}^r \mathbb{N}_i(t) + \hat{B}_\Delta(t)$. Compute the new budget allocation, $\mathbb{N}_1', \mathbb{N}_2', \ldots$, and $\mathbb{N}_r'$ via (6.5) $\sim$ (6.6).

2: Perform additional $max(0, \mathbb{N}_i' - \mathbb{N}_i(t))$ for action i, $\forall i \in \mathbb{Z}_r$. For each action i, when it consumes one budget to interact with the environment and receives a feedback $\beta_i(t) \in \{0, 1\}$, update $\tilde{d}_i(t)$ according to the following equations for the chosen action

$$\mathbb{H}_i(t+1) = \mathbb{H}_i(t) + \beta_i(t),$$

$$\mathbb{N}_i(t+1) = \mathbb{N}_i(t) + 1,$$

$$\tilde{d}_i(t+1) = \mathbb{H}_i(t+1)/\mathbb{N}_i(t+1),$$

$$t = t + 1,$$

3: Update $P(t)$ according to the update scheme of LA$_{OCBA}$ in (6.8) $\sim$ (6.9). Update $\hat{B}_\Delta(t)$ according to (6.7).

4: If $\max_{i \in \mathbb{Z}_r} \{p_i(t)\} \geq \hat{T}$, **Then** CONVERGE to the action m whose $\tilde{d}_m(t) = \max\{\tilde{d}_i(t)\}$. **Else** *Goto* Step 1. **Endif**

END

In traditional LA, $P(t)$ is updated according to the response from the environment. While in LA$_{OCBA}$, a cost-effective sequential approach based on OCBA is employed to accomplish the ordinal optimization. The sampling budget is now allocated to actions in a way that maximizes the probability of selecting the optimal action correctly. From Steps 1–3, $P(t)$ is updated only when the additional budget $\hat{B}_A(t)$ is used up, i.e., a **batch** of responses have been received.

Example 6.2 Let $D = \{0.75, 0.8, 0.6\}$ in dictating the true optimal action is α_2. After initial sampling $\hat{B}_0$ times, $\tilde{D} = \{0.75, 0.7, 0.6\}$. An **adaptive** $\hat{B}_A(t)$ is allocated to actions through OCBA in Step 1 and used for sampling in Step 2. The Reward–Penalty [1] update scheme is adopted to update $P(t)$ as same as in Example 6.1. When $P = \{0, 1, 0\}$, the algorithm terminates. This example shows that OCBA is utilized to optimize the allocation of sampling budget to maximize the probability of identifying the true optimal action. P is used as a stop condition which actually determines the overall budget of OCBA.

The additional sampling budget $\hat{B}_A(t)$ decreases with the $p_m(t)$ till $\hat{B}_A(t) = 1$, i.e.,

$$\hat{B}_A(t) = \max\{1, \text{round}(\hat{B}_0(1 - p_m(t)))\}, \tag{6.7}$$

where round(x) means to round x to the nearest integer, e.g., round(1.1) = 1 and round(1.8) = 2. $\hat{B}_A(1)$ is initialized by $\hat{B}_0$. The $p_m(t)$ is the probability of the estimated optimal action. As $p_m(t)$ increases, $\hat{B}_A(t)$ decreases to speed up the update of $P(t)$ such that the convergence speed is promoted. This is reasonable since the estimate of the optimal action turns to be more and more stable and accurate as $p_m(t)$ increases. Therefore, the action probability vector $P(t)$ is updated in an **adaptive batch** way. In Step 3, the Reward–Penalty [1] update scheme is adopted to update $P(t)$. As a batch of sampling is finished, let $\bar{c}_t \in \{1, 2, \ldots, r-1\}$ be the number of all the actions that are not the estimated optimal action and decrease their action probabilities by a step $\Delta = \frac{1}{n}$, where r is the number of allowable actions and n is a resolution parameter. Thus, $\bar{c}_t \Delta$ is an integral multiple of Δ according to the action probability updating rule in (6.8). It is notable that the action whose probability is already 0 does not decrease its probability any more. In order to keep the sum of the components of $P(t)$ equal to one, the probability of the estimated optimal action has to be increased by $\bar{c}_t \Delta$ according to (6.9). The update scheme of $P(t)$ is as follows.

$$p_j(t+1) = \max\{p_j(t) - \Delta, 0\}, \quad j \neq m, \quad \forall j \in \mathbb{Z}_r, \tag{6.8}$$

$$p_m(t+1) = 1 - \Sigma_{j \neq m} p_j(t+1). \tag{6.9}$$

Step 1 will then continue until $\max_{i \in \mathbb{Z}_r}\{p_i(t)\} \geq \hat{T}$. In LA_{OCBA} (Step 4), $P(t)$ only determines when to terminate the algorithm and hence the size of sampling budget. The task of how to allocate the sampling budget among actions to maximize the probability of correctly selecting the optimal action is separated from $P(t)$ and handed over to OCBA. In this framework, $P(t)$ and OCBA take charge of the size of sampling budget and how to allocate the sampling budget among actions, respectively, in an independent way.

6.3 Proof of Optimality

It is critically important to investigate if an learning automaton is ε-optimal. Recently, a common flaw was discovered by the authors of [17] in the previous proofs for the ε-optimality of almost all of the reported estimator algorithms. A new proof method that fixes this flaw is introduced in [17] by invoking the monotonicity property. Another proof method is then introduced in [18–20] to fix the flaw in the proof for the ε-optimality of the pursuit learning automaton. But instead of invoking the monotonicity property of the action probabilities, it examines their submartingale property and invokes the theory of Regular functions to prove the ε-optimality. The latter one has been adopted in [21] and is also adopted here to prove the ε-optimality of our introduced algorithm.

The learning process of LA_{OCBA} can be divided into two phases: The first one is the sufficient sampling for all actions in order to make the optimal action's rank accurate with the probability equal to or greater than $1 - \delta$ given a sufficiently large value of the learning parameter n. The second phase is to increase the probability of the estimated optimal action such that it converges to one. Because the true optimal action ranks first with high accuracy in the second phase, $p_m(t)$ keeps increasing such that the optimal action will not be missed with probability $1 - \delta$.

Theorem 6.1 *For any given constants $\delta > 0$ and $M < \infty$, there exist $n_0 < \infty$ and $t_0 < \infty$ such that under LA_{OCBA}, for all learning parameters $n > n_0$ and all time $t > t_0$:*

$$\Pr\{\mathbb{N}_i(t) \geq M\} \geq 1 - \delta, \quad i \in \mathbb{Z}_r,$$

where $\mathbb{N}_i(t)$ is the number of times the ith action is chosen up to time t in any specific realization.

Proof: Consider LA_{OCBA}, for an action, say α_i, we will give a threshold of the resolution parameter, denoted by $n(i)$, by which $p_i(t)$ remains non-zero if $t \leq t(i)$. Here, $t(i)$ is a time instant value, which will be defined later.

For action α_i, by virtue of LA_{OCBA}, at time $t \leq t(i)$:

$$
\begin{aligned}
p_i(t) &\geq \left(p_i(0) - \sum_{j=1\ldots t(i)} \frac{1}{rn(i)} \right) \\
&\geq \left(p_i(0) - \sum_{j=1\ldots t(i)} \frac{1}{2n(i)} \right) \\
&= \left(p_i(0) - \frac{t(i)}{2n(i)} \right), \quad i \in \mathbb{Z}_r.
\end{aligned}
\tag{6.10}
$$

Then, under the premise that $(p_i(0) - t(i)/2n(i)) > 0$, we can arbitrarily set the value of $n(i)$ to be

$$n(i) = r \cdot t(i) \tag{6.11}$$

such that LA_{OCBA} will not converge before $t(i)$.

Second, we investigate the existence of $t(i)$, by which for any given constants $\delta > 0$ and $M < \infty$, when $t > t(i)$ and all learning parameters $n > n(i) = r \cdot t(i)$, it follows that $\Pr\{\mathbb{N}_i(t) \geq M\} \geq 1 - \delta$.

For all actions α_i with $i \neq m$: Let w denote the difference between the two highest reward probabilities and $\mathbb{N}(t) = t(i)$. From (6.6), we have

$$\mathbb{N}_i(t(i)) = \frac{t(i)\frac{\hat{\sigma}_i^2(t)}{\hat{\delta}_{i,m}^2(t)}}{\hat{\sigma}_m(t)\sum_{j=1,j\neq m}^r \frac{\hat{\sigma}_j(t)}{\hat{\delta}_{j,m}^2(t)} + \sum_{j=1,j\neq m}^r \frac{\hat{\sigma}_j^2(t)}{\hat{\delta}_{j,m}^2(t)}}$$

$$\geq \frac{t(i)\frac{\hat{\sigma}_i^2(t)}{\hat{\delta}_{i,m}^2(t)}}{\hat{\sigma}_m(t)\sum_{j=1,j\neq m}^r \frac{\hat{\sigma}_j(t)}{w^2} + \sum_{j=1,j\neq m}^r \frac{\hat{\sigma}_j^2(t)}{w^2}}$$

$$\geq \frac{t(i)\frac{\hat{\sigma}_i^2(t)}{\hat{\delta}_{i,m}^2(t)}}{0.25\sum_{j=1,j\neq m}^r \frac{0.25}{w^2} + \sum_{j=1,j\neq m}^r \frac{0.25^2(t)}{w^2}}$$

$$\geq \frac{8w^2 t(i)\frac{\hat{\sigma}_i^2(t)}{\hat{\delta}_{i,m}^2(t)}}{(r-1)}$$

$$\geq \frac{8w^2 t(i)\hat{\sigma}_i^2(t)}{(r-1)}, \tag{6.12}$$

where constant 0.25 is based on the fact that the Bernoulli standard deviation ≤ 0.25.

Then, we have

$$\Pr\{\mathbb{N}_i(t) \geq M\} \geq \Pr\left\{\frac{8w^2 t(i)\hat{\sigma}_i^2(t)}{(r-1)} \geq M\right\}$$

$$= \Pr\left\{\hat{\sigma}_i^2(t) \geq \frac{M(r-1)}{8w^2 t(i)}\right\}. \tag{6.13}$$

According to (6.11), we have

$$\Pr\{\mathbb{N}_i(t) \geq M\} \geq \Pr\left\{\hat{\sigma}_i^2(t) \geq \frac{r(r-1)M}{8w^2 n(i)}\right\}. \tag{6.14}$$

To bound the right side of (6.14), we introduce the following argument: Since $0 < d_i < 1$, it holds that

$$\Pr\left\{\hat{\sigma}_i^2(t) = 0\right\} \le d_i^{\hat{B}_0} + (1 - d_i)^{\hat{B}_0} \le \frac{1}{2^{\hat{B}_0 - 1}},$$

where $\hat{B}_0$ is defined in Section 6.2.

For any given $\delta > 0$, when $\hat{B}_0 > \lceil 1 - \log_2 \delta \rceil$, there exists a constant $\epsilon > 0$ such that

$$\Pr\left\{\hat{\sigma}_i^2(t) \ge \epsilon\right\} > 1 - \delta.$$

Then, when $n(i) > \lceil \frac{r(r-1)M}{8w^2\epsilon} \rceil$, the right side of (6.14) follows that

$$\Pr\left\{\hat{\sigma}_i^2(t) \ge \frac{r(r-1)M}{8w^2 n(i)}\right\} > 1 - \delta.$$

Therefore, we can obtain

$$\Pr\left\{\mathbb{N}_i(t) \ge M\right\} > 1 - \delta. \tag{6.15}$$

Hence, we can repeat this argument for all the actions, and we can finally define t_0 and n_0 as follows:

$$t_0 = \max_{i \in \mathbb{Z}_r}\{t(i)\}, \tag{6.16}$$

$$n_0 = \max_{i \in \mathbb{Z}_r}\{n(i)\} = \max_{i \in \mathbb{Z}_r}\{r \cdot t(i)\}, \tag{6.17}$$

which can complete the proof for all actions α_i with $i \neq m$.

For action α_m: Using a similar procedure by replacing (6.6) with (6.5), we can prove the result for action α_m.

Combining the proofs for the two cases above, we have the conclusion. ∎

The proof of this theorem implies that under LA_{OCBA}, by utilizing a sufficiently large value of learning parameter n_0, each action is to be selected for an arbitrarily large number of times. Based on Theorem 6.1, we can further establish the following theorem.

Theorem 6.2 *Given a $\delta \in (0, 1)$, there exists a time instant $t_0 < \infty$, such that $\Pr\{\overline{B}(t_0)\} = 1$ where $\overline{B}(t_0)$ is defined as follows:*

$$\overline{q}_j(t) = \Pr\{|\tilde{d}_j(t) - d_j| < \frac{w}{2}\}, \tag{6.18}$$

$$\overline{q}(t) = \Pr\{|\tilde{d}_j(t) - d_j| < \frac{w}{2}, \quad \forall j \in (1, 2, \ldots, r)\}$$

$$= \prod_{j=1,2,\ldots,r} \overline{q}_j(t), \tag{6.19}$$

$$\tilde{B}(t) = \{\overline{q}(t) > 1 - \delta\}, \quad \delta \in (0, 1), \tag{6.20}$$

$$\overline{B}(t_0) = \{\bigcap_{t>t_0}\{\overline{q}(t) > 1 - \delta\}\}, \quad \delta \in (0, 1), \tag{6.21}$$

where w is the difference between two highest reward probabilities.

This theorem gives the key condition, i.e., $\overline{B}(t_0)$ for $p_m(t)_{t>t_0}$ with $t_0 < \infty$ being a submartingale. Hence, the flaw in the previous proofs reported in the literature where $\tilde{B}(t)$ is equivalent to $\overline{B}(t_0)$ can be eliminated. This theorem's proof can be found in [18–20].

Theorem 6.3 *Under* LA_{OCBA}, *there exists* $t_0 < \infty$ *such that* $p_m(t)_{t>t_0}$ *is a submartingale.*

Proof: Recall ALGORITHM LA_{OCBA} where Reward–Penalty scheme is used as the updating rule. From Theorem 6.1, for all $t \geq t_0$, it follows that:

$$E[p_m(t+1)|Q(t)]$$
$$\geq (\overline{q}(t)(p_m(t) + \overline{c}_t\Delta) + (1 - \overline{q}(t))(p_m(t) - \Delta) \tag{6.22}$$
$$= ((\overline{c}_t + 1)\overline{q}(t) - 1)\Delta + p_m(t),$$

where $\overline{c}_t\Delta$ is an integral multiple of Δ and $\overline{c}_t = 1, 2, \ldots, r - 1$ according to the action probability updating rules of LA_{OCBA}.

Then,

$$E[p_m(t+1)|Q(t)] - p_m(t) \geq (\overline{c}_t q(t) - 1)\Delta. \tag{6.23}$$

If we set $1 - \delta$ defined in Theorem 6.2 to be greater than or equal to $\frac{1}{2}$, then for every single time instant subsequent to $t > t_0, \overline{q}(t) \geq 1 - \delta \geq \frac{1}{2}$,

$$E[p_m(t+1) - p_m(t)|Q(t)] \geq 0, \tag{6.24}$$

which implies that $p_m(t)_{t>t_0}$ is a submartingale. ∎

Based on Theorem 6.3, by using the Martingale convergence theory [22], we can derive the following corollary.

Corollary 6.1 *Under* LA_{OCBA},

$$\lim_{t\to\infty} p_m(t) = 0 \ or \ 1.$$

Theorem 6.4 LA_{OCBA} *is ε-optimal in all stationary environments. Formally, let* $\hat{T} = 1 - \varepsilon$ *be a value arbitrarily close to 1, with ε being arbitrarily small. Then, given any δ, there exists a positive integer $n_0 < \infty$ and a time instant $t_0 < \infty$, such that for all learning parameters $n > n_0$ and for all $t > t_0$,* $\Pr\{\lim_{t\to\infty} p_m(t) = 1\} = 1$.

Proof: Denoting e_j as the unit vector with the jth element being 1, we need to prove the convergence probability

$$\Gamma_m(P) = \Pr\{\lim_{t\to\infty} p_m(t) = 1 | P(0) = P\}$$
$$= \Pr\{\lim_{t\to\infty} P(t) = e_m | P(0) = P\} = 1. \tag{6.25}$$

Define a function $\Phi_m(P) = e^{-\tilde{x}_m P_m}$, where $\tilde{x}_m$ is a positive constant and then define an operator $\bar{U}$ as:

$$\bar{U}(\Phi_m(P)) = E[\Phi_m(P(t+1)) | P(t) = P]. \tag{6.26}$$

Under LA_{OCBA}, it follows that:

$$\bar{U}(\Phi_m(P)) - \Phi_m(P)$$
$$= E[\Phi_m(P(t+1)) | P(t) = P] - \Phi_m(P)$$
$$= (\overline{q}(t)(e^{-\tilde{x}_m(p_m(t)+\overline{c}_t\Delta)})$$
$$+ (1 - \overline{q}(t))(e^{-\tilde{x}_m(p_m(t)-\Delta)})) - e^{-\tilde{x}_m P_m}$$
$$= \overline{q}(t)(e^{-\tilde{x}_m(p_m(t)+\overline{c}_t\Delta)} - e^{-\tilde{x}_m(p_m(t)-\Delta)})$$
$$+ e^{-\tilde{x}_m(p_m(t)-\Delta)} - e^{-\tilde{x}_m P_m}$$
$$= e^{-\tilde{x}_m p_m(t)}(\overline{q}(t)(e^{-\overline{c}_t\tilde{x}_m\Delta} - e^{\tilde{x}_m\Delta})$$
$$+ (e^{\tilde{x}_m\Delta} - 1)). \tag{6.27}$$

A proper value should be determined for $\tilde{x}_m$ such that $\bar{U}(\Phi_m(P)) - \Phi_m(P) \leq 0$. This is equivalent to solving the following inequality:

$$\overline{q}(t)(e^{-\overline{c}_t\tilde{x}_m\Delta} - e^{\tilde{x}_m\Delta}) + (e^{\tilde{x}_m\Delta} - 1) \leq 0. \tag{6.28}$$

We know that when $\tilde{b} > 0$ and $x \to 0$, $\tilde{b}^x = 1 + (\ln\tilde{b})x + \frac{((\ln\tilde{b})^2)}{2}x^2$. If we set $\tilde{b} = e^{-\tilde{x}_m}$, when $\Delta \to 0$, (6.28) can be re-written as

$$\overline{q}(t)\left((\ln\tilde{b})(\overline{c}_t + 1)\Delta + \frac{(\ln\tilde{b})^2}{2}(\overline{c}_t^2 - 1)\Delta^2\right)$$
$$- (\ln\tilde{b})\Delta + \frac{(\ln\tilde{b})^2}{2}\Delta^2 \leq 0. \tag{6.29}$$

Substituting $\tilde{b}$ with $e^{-\tilde{x}_m}$, we have

$$\tilde{x}_m\left(\tilde{x}_m - \frac{2(\overline{q}(t)(\overline{c}_t + 1) - 1)}{\Delta(\overline{q}(t)(\overline{c}_t^2 - 1) + 1)}\right) \leq 0. \tag{6.30}$$

As $\tilde{x}_m$ is defined as a positive constant, we have

$$0 < \tilde{x}_m \leq \frac{2(\overline{q}(t)(\overline{c}_t + 1) - 1)}{\Delta(\overline{q}(t)(\overline{c}_t^2 - 1) + 1)}. \tag{6.31}$$

Denote $\tilde{x}_{m0} = \frac{2(\bar{q}(t)(\bar{c}_t+1)-1)}{\Delta(\bar{q}(t)(\bar{c}_t^2-1)+1)}$. When $\Delta \to 0$, $\tilde{x}_{m0} \to \infty$ as $\bar{c}_t = 1, 2, \ldots, r-1$ and $\bar{q}(t)_{t>t_0} > \frac{1}{2}$.

Thus, $\Phi_m(P)$ is superregular according to the following definitions of subregular/superregular functions in [22]:

A function $\chi \in C(\hat{S}_r)$ is called *superregular* if

$$\chi(p) \geq \bar{U}(\chi(p)), \quad \forall p \in \hat{S}_r, \tag{6.32}$$

where $C(\hat{S}_r)$ denotes the class of all continuous functions mapping $\hat{S}_r \to \mathfrak{R}$ and $\mathfrak{R}$ is the real line.

Similarly, a function $\chi(\cdot)$ is called *regular* if

$$\chi(p) = \bar{U}(\chi(p)) \tag{6.33}$$

and *subregular*, if

$$\chi(p) \leq \bar{U}(\chi(p)) \tag{6.34}$$

over $p \in \hat{S}_r$.

Let $\phi(P) = \frac{1-e^{-\tilde{x}_m P_m(t)}}{1-e^{-\tilde{x}_m}}$, where $\tilde{x}_m$ is the same as defined in $\chi(p)$. $\phi(P)$ also meets the boundary conditions, i.e., $0 \leq \phi(P) \leq 1$ obviously.

Moreover, $\phi(P)$ is a subregular (superregular) if $\Phi_m(P)$ is a superregular (subregular) [22]. Therefore, (6.31) ensures that $\Phi_m(P)$ is superregular, which leads $\phi(P)$ to be subregular.

According to the theory of regular functions [22], we have

$$\Gamma_m(P) \geq \phi(P) = \frac{1 - e^{-\tilde{x}_m P_m(t)}}{1 - e^{-\tilde{x}_m}}, \tag{6.35}$$

which meets the boundary conditions of (6.25). As (6.35) holds for every $\tilde{x}_m$ bounded by (6.31), if we take the greatest value of $\tilde{x}_{m0}$, as $\tilde{x}_{m0} \to \infty$, such that $\Gamma_m(P) \to 1$ according to the definition of $\Gamma_m(P)$ in (6.25). We have thus proved that $\Pr\{\lim_{t\to\infty} P_m(t) = 1\} = 1$, showing that LA$_{OCBA}$ is ε-optimal. ∎

6.4 Simulation Studies

Can the use of OCBA in LA bring about significant advantages? This section presents the experimental results comparing LA_{OCBA} with LELA [6], LACP [7], DP$_{RI}$ [1, 3], and DGPA [4] in terms of their accuracy and convergence speed. LELA$_S$ and LELA$_R$ are two variants of LELA using two strategies, respectively, to choose one from the active actions with identical minimal estimated reward probability as the last-position action [6]. LELA$_S$ returns the one with the smallest action index, while LELA$_R$ chooses it randomly. LELA$_R$ is chosen in this simulation to represent LELA since more significant results are

Table 6.1 Reward probability.

	d_1	d_2	d_3	d_4	d_5	d_6	d_7	d_8	d_9	d_{10}
E_1	**0.65**	0.50	0.45	0.40	0.35	0.30	0.25	0.20	0.15	0.10
E_2	**0.60**	0.50	0.45	0.40	0.35	0.30	0.25	0.20	0.15	0.10
E_3	**0.55**	0.50	0.45	0.40	0.35	0.30	0.25	0.20	0.15	0.10
E_4	**0.70**	0.50	0.30	0.20	0.40	0.50	0.40	0.30	0.50	0.20
E_5	0.10	0.45	**0.84**	0.76	0.20	0.40	0.60	0.70	0.50	0.30

The bold value means the max reward.

obtained with it [6]. For LACP, $\theta_1 = 0.2$ is the best one obtained on the benchmark environments. $\theta_2 = 0$ as suggested in [7].

The popular benchmark test sets with ten actions presented in Table 6.1 are used to examine the performances of contenders. E_1–E_5 are popular benchmarks and were used in [1, 3–6, 23]. Each environment includes ten allowable actions whose probabilities for each environment are presented in Table 6.1. An algorithm is considered to have converged if the probability of choosing an action is not less than a threshold $\hat{T}$ ($0 < \hat{T} \leq 1$) in all performed tests. If the automaton converges to the action that has the highest reward probability, it is considered to have converged correctly.

As a common sense, the step size is a critical trade-off between the convergence accuracy and speed. Accuracy is defined as the ratio between the number of correctly converged ones and the total number of experiments. Convergence speed is measured by the minimal average number of iterations required for the same given accuracy equals to 99.5%. Bigger resolution parameter n leads to slower convergence and vice versa. To compare the convergence speed of the contenders in a fair way, the minimum value of n is considered as the best value if it yields the fastest convergence given the accuracy equals to 99.5%. Each result is obtained from 50 000 runs for each algorithm in each environment. These best parameters are then chosen as the final parameter values used for the contenders to compare their performances. Each algorithm first samples all actions ten times each in order to initialize the estimated reward probability vector. These extra iterations are also included in the results presented in Table 6.3. Let $\hat{T} = 0.999$, each performance result of contenders is obtained from 250 000 runs for each algorithm in each environment by using their best learning parameters.

For the real applications, an automatic parameter tuning method can be adopted: The parameter values are considered as the best values if they yield the fastest convergence and the automaton converges to the correct action in a sequence of $\mathbb{N}_E$ experiments. It is recommended that $\mathbb{N}_E = 750$ in [1, 3, 4, 24].

Table 6.2 The convergence speed of LA$_{OCBA}$ using different $\hat{B}_0$ ("ϱ" denotes the best parameter of contenders and "ϑ" denotes the convergence speed).

	$\hat{B}_0 = 100$		$\hat{B}_0 = 50$		$\hat{B}_0 = 20$		$\hat{B}_0 = 10$	
	ϱ	ϑ	ϱ	ϑ	ϱ	ϑ	ϱ	ϑ
E_1	7	526	13	480	31	453	60	**451**
E_2	15	994	29	994	70	915	136	**914**
E_3	55	3261	111	3254	273	**3235**	550	3343
E_4	6	470	10	402	24	379	47	**376**
E_5	13	885	26	851	64	829	123	**822**

The bold value means the fastest convergence speed.

Initially, $\hat{B}_0$ samples for each action are conducted to obtain some information about the performance of each action during the first stage. As the simulation proceeds, the sample mean and variance of each action are computed from the data already collected up to that stage. According to this observed simulation output, a sequentially incremental budget $\hat{B}_\Delta$ is allocated based on (6.1) and (6.2) and the simulation iterates.

A large $\hat{B}_0$ can result in the waste of computation time to obtain an unnecessarily high confidence level. On the other hand, if $\hat{B}_0$ is too small, we need to compute (6.1) and (6.2) many times [10, 11]. In the same time, small $\hat{B}_0$ also makes more frequent update of the action probability vector. Table 6.2 shows the convergence speed of LA$_{OCBA}$ using different $\hat{B}_0$. It can be seen that small $\hat{B}_0$ leads to faster convergence speed in most cases. However, this speedup turns to be inefficient when $\hat{B}_0 < 20$. Therefore, $\hat{B}_0 = 20$ is adopted in the following comparisons, which is also suggested in some other applications.

Table 6.3 The convergence speed of contenders ("ϱ" denotes the best parameter of contenders and "ϑ" denotes the convergence speed).

	LA$_{OCBA}$		LELA$_R$		DGPA		LACP		DP$_{RI}$	
	ϱ	ϑ	ϱ	ϑ	ϱ	ϑ	ϱ	ϑ	ϱ	ϑ
E_1	31	**453**	8	573	27	748	250	974	298	1088
E_2	70	**915**	17	1129	59	1543	604	2385	679	2562
E_3	273	**3325**	59	3733	198	5059	3510	11 809	3801	12 430
E_4	24	**379**	8	534	23	646	154	650	205	767
E_5	64	**829**	22	998	48	1287	844	2305	1001	2579

The bold value means the fastest convergence speed.

Table 6.4 Convergence speedup rate of LA_{OCBA} compared with contenders.

	LELA$_R$ (%)	DGPA (%)	LACP (%)	DP$_{RI}$ (%)
E_1	20.94	39.44	53.49	42.09
E_2	18.95	40.70	61.64	50.08
E_3	10.93	34.28	71.84	65.94
E_4	29.03	41.33	41.69	25.82
E_5	16.93	35.59	64.03	52.75

The convergence speed of the contenders is presented in Table 6.3. It is shown that LA_{OCBA} has faster convergence speed than $LELA_R$, LACP, DP_{RI}, and DGPA in all of the benchmark environments. In Table 6.3, "ϱ" denotes the best parameter of contenders and "ϑ" denotes the convergence speed which is the mean number of iterations needed for convergence.

The speedup rates of LA_{OCBA} compared with DP_{RI}, LACP, DGPA, and $LELA_R$ are presented in Table 6.4 where the speedup rate is computed as $(\vartheta_{contender} - \vartheta_{LA_{OCBA}})/\vartheta_{contender}$, where ϑ means the average number of iterations. For example, the speedup rate compared with DP_{RI} in E_1 is computed as $(\vartheta_{DP_{RI}} - \vartheta_{LA_{OCBA}})/\vartheta_{DP_{RI}}$. These speedup results are very significant ranging from 10.93% to 65.94%.

In addition to the numerical results, the convergence curves of LA_{OCBA} and the contenders are illustrated in Fig. 6.2. They show the mean choice probability $p_m(t)$ of the optimal action as a function of the iteration number. The higher $p_m(t)$, the faster convergence to the optimal action because the LA algorithm is considered to have converged if the probability of choosing an action is not less than a threshold $\hat{T}$ in all the performed tests. Therefore, LA_{OCBA} converges faster than the classical pursuit algorithms DP_{RI}, DGPA, LACP, and $LELA_R$.

Let $\hat{\alpha}_m(t) = \mathbb{N}_m(t)/\mathbb{N}(t)$ under LA_{OCBA}, which means the sampling rate allocated to the estimated optimal action m at time t. For DP_{RI}, LACP, DGPA, and $LELA_R$, the sampling rate $\hat{\alpha}_m(t)$ is exactly equal to the choice probability $p_m(t)$. Figure 6.3 illustrates the mean sampling rate curves of the estimated optimal action among contenders to characterize their essential differences and verify the motivation of our chapter.

Figure 6.3 shows that, under LA_{OCBA}, $\hat{\alpha}_m(t)$ begins to be larger in the initial phase due to the importance of the estimated optimal action m and then decreases slowly with the convergence because the sample variance of the estimated optimal action m turns to be small. In this way, compared with the classical LA, the estimated optimal action m is sampled more in the initial stage and less in the later convergent stage. Sampling is allocated to actions in a way that maximizes the probability

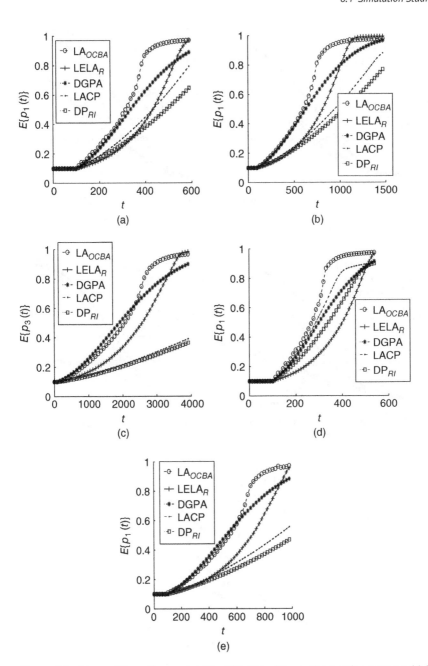

Figure 6.2 Mean of the action probability $E\{P_m(t)\}$ of the optimal action versus t. (a) E_1, (b) E_2, (c) E_3, (d) E_4, and (e) E_5.

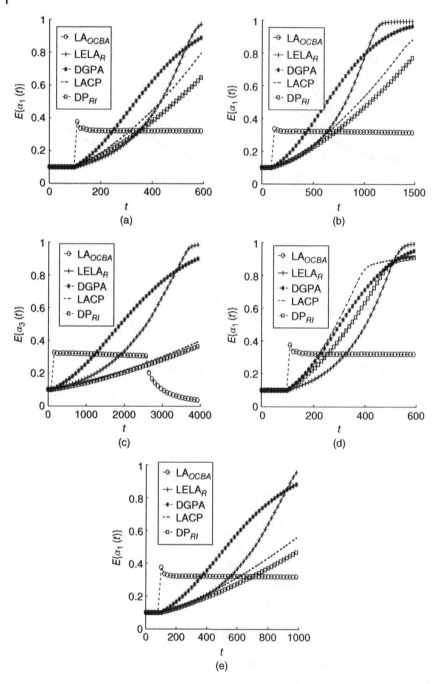

Figure 6.3 Mean of the $E\{\hat{\alpha}_m(t)\}$ versus t. (a) E_1, (b) E_2, (c) E_3, (d) E_4, and (e) E_5.

of selecting the optimal action other than the traditional way through the action probability vector $P(t)$.

6.5 Summary

Learning automata (LAs) have gained many applications [25–46] as a powerful reinforcement learning method. This chapter introduces a new class of LAs called LA_{OCBA} by using OCBA for the ordinal optimization such that the probability of selecting the optimal action correctly is maximized. Its ε-optimality is proved. Simulation results verify its significant advantages over the state-of-the-art methods in terms of convergence speed and learning accuracy. Its applications to various systems and networks [41–46] remain to be explored.

6.6 Exercises

1 What are the roles of action probability vector in learning automata?

2 What is the relationship between learning automata and ordinal optimization?

3 What is the main idea of OCBA?

4 Which components does a learning model have? Describe their functionalities.

5 Design an instance where the traditional pursuit scheme in learning automata fails to identify the optimal action while LA_{OCBA} can.

6 Run LA_{OCBA} with different parameters $\hat{B}_0$ and n to test the sensitivity of its performance versus the parameter settings.

7 When OCBA is incorporated into LA, should the overall budget of LA be predefined? Justify your answer.

8 What are the respective roles of OCBA and LA in LA_{OCBA}?

9 Give an example of problems that LA can solve but OCBA cannot.

10 Describe the flow of LA_{OCBA}.

11 Design a twenty-action set in a noisy environment and use OCBA, LA, and LA_{OCBA} to learn the optimal action. Then compare their performance.

References

1 B. J. Oommen and M. Agache, "Continuous and discretized pursuit learning schemes: Various algorithms and their comparison," *IEEE Transactions on Systems, Man, and Cybernetics, Part B (Cybernetics)*, vol. 31, no. 3, pp. 277–287, June 2001.

2 M. A. L. Thathachar and P. S. Sastry, "A class of rapidly converging algorithms for learning automata," in *Proceedings of the IEEE International Conference on Cybernetics and Society*, Bombay, India, 1984, pp. 602–606.

3 B. J. Oommen and J. K. Lanctot, "Discretized pursuit learning automata," *IEEE Transactions on Systems, Man, and Cybernetics*, vol. 20, pp. 931–938, July/Aug 1990.

4 M. Agache and B. J. Oommen, "Generalized pursuit learning schemes: New families of continuous and discretized learning automata," *IEEE Transactions on Systems, Man, and Cybernetics, Part B (Cybernetics)*, vol. 32, no. 6, pp. 738–749, 2002.

5 M. S. Georgios I. Papadimitriou, and A. S. Pomportsis, "A new class of ε-optimal learning automata," *IEEE Transactions on Systems, Man, and Cybernetics, Part B (Cybernetics)*, vol. 34, no. 1, pp. 246–254, 2004.

6 J. Zhang, C. Wang, and M. Zhou, "Last-position elimination-based learning automata," *IEEE Transactions on Cybernetics*, vol. 44, no. 12, pp. 2484–2492, 2014.

7 E. Ikonen, "Optimization using learning automata and confidence probabilistics," in *IFAC International Conference on Intelligent Control Systems and Signal Processing*, Faro, Portugal, 2003, pp. 197–202.

8 E. Ikonen and K. Najim, "Online optimization of replacement policies using learning automata," *International Journal of Systems Science*, vol. 39, no. 3, pp. 237–249, 2008.

9 Y. C. Ho, R. S. Sreeniva, and P. Vakili, "Ordinal optimization of DEDS," *Discrete Event Dynamic Systems*, vol. 2, no. 2, pp. 61–88, 1992.

10 C.-H. Chen, S. D. Wu, and L. Dai, "Ordinal comparison of heuristic algorithms using stochastic optimization," *IEEE Transactions on Robotics and Automation*, vol. 15, no. 1, pp. 44–56, 1999.

11 H.-C. Chen, C.-H. Chen, and E. Yücesan, "Computing efforts allocation for ordinal optimization and discrete event simulation," *IEEE Transactions on Automatic Control*, vol. 45, no. 5, pp. 960–964, 2000.

12 C.-H. Chen, J. Lin, E. Yücesan, and S. E. Chick, "Simulation budget allocation for further enhancing the efficiency of ordinal optimization," *Discrete Event Dynamic Systems: Theory and Applications*, vol. 10, pp. 251–270, 2000.

13 C.-h. Chen and L. H. Lee, "Stochastic simulation optimization: An optimal computing budget allocation," World Scientific, 2010.

14 C. Y. P. A. Johnson and N. Lin, "Source code of optimal computing budget allocation (OCBA)," http://www.ise.nus.edu.sg/staff/leelh/research/OCBA/demo.html.

15 A. M. Turing, "Computing machinery and intelligence," *Mind*, vol. 59, pp. 433–460, 1950.

16 R. Stuart and N. Peter, "Artificial intelligence: A modern approach," Prentice Hall, 2002.

17 M. Ryan and T. Omkar, "On ε-optimality of the pursuit learning algorithm," *Journal of Applied Probability*, vol. 49, no. 3, pp. 795–805, 2012.

18 X. Zhang, O.-C. Granmo, B. J. Oommen, and L. Jiao, "On using the theory of regular functions to prove the ε-optimality of the continuous pursuit learning automaton," in *Proceedings of the International Conference on Industrial and Engineering Applications of Artificial Intelligence and Expert Systems*, June 2013, pp. 262–271.

19 X. Zhang, O.-C. Granmo, B. J. Oommen, and L. Jiao, "A formal proof of the ε-optimality of absorbing continuous pursuit algorithms using the theory of regular functions," *Applied Intelligence*, vol. 41, pp. 974–985, 2014.

20 X. Zhang, B. J. Oommen, O.-C. Granmo, and L. Jiao, "Using the theory of regular functions to formally prove the ε-optimality of discretized pursuit learning algorithms," in *Modern Advances in Applied Intelligence*. Springer, 2014, pp. 379–388.

21 J. Zhang, C. Wang, and M. Zhou, "Fast and epsilon-optimal discretized pursuit learning automata," *IEEE Transactions on Cybernetics*, vol. 45, no. 10, pp. 2089–2099, 2015.

22 K. S. Narendra and M. A. L. Thathachar, "Learning automata: An introduction," Englewood Cliffs, NJ: Prentice-Hall, 1989.

23 M. A. L. Thathachar and P. S. Sastry, "A new approach to the design of reinforcement schemes for learning automata," *IEEE Transactions on Systems, Man, and Cybernetics*, vol. SMC-15, pp. 168–175, Jan/Feb 1985.

24 M. A. L. Thathachar and B. J. Oommen, "Discretized reward-inaction learning automata," *Journal of Cybernetics and Information Science*, vol. 2, pp. 24–29, 1979.

25 B. J. Oommen and M. K. Hashem, "Modeling a student's behavior in a tutorial-like system using learning automata," *IEEE Transactions on Systems, Man, and Cybernetics, Part B (Cybernetics)*, vol. 40, no. 2, pp. 481–492, 2009.

26 B. J. Oommen and M. K. Hashem, "Modeling a student-classroom interaction in tutorial-like system using learning automata," *IEEE Transactions on Systems, Man, and Cybernetics, Part B (Cybernetics)*, vol. 40, no. 1, pp. 29–42, 2010.

27 B. J. Oommen and M. K. Hashem, "Modeling the learning process of the teacher in a tutorial-like system using learning automata," *IEEE Transactions on Cybernetics*, vol. 43, no. 6, pp. 2020–2031, 2013.

28 Y. Wang, W. Jiang, Y. Ma, H. Ge, and Y. Jing, "Learning automata based cooperative student-team in tutorial-like system," in *Intelligent Computing Methodologies*. Springer, 2014, pp. 154–161.

29 A. V. Melkikh, "Quantum information and the problem of mechanisms of biological evolution," *Biosystems*, vol. 115, pp. 33–45, 2014.

30 B. Masoumia and M. Meybodi, "Speeding up learning automata based multi agent systems using the concepts of stigmergy and entropy," *Expert Systems with Applications*, vol. 38, no. 7, pp. 8105–8118, 2011.

31 D. Maravall, J. De Lope, and J. P. Fuentes, "Fusion of probabilistic knowledge-based classification rules and learning automata for automatic recognition of digital images," *Pattern Recognition Letters*, vol. 34, no. 14, pp. 1719–1724, 2013.

32 E. Cuevas, F. Wario, D. Zaldivar, and M. Pérez-Cisneros, "Circle detection on images using learning automata," in *Artificial Intelligence, Evolutionary Computing and Metaheuristics*. Springer, 2013, pp. 545–570.

33 C. Unsal, P. Kachroo, and J. S. Bay, "Multiple stochastic learning automata for vehicle path control in an automated highway system," *IEEE Transactions on Systems, Man, and Cybernetics - Part A: Systems and Humans*, vol. 29, no. 1, pp. 120–128, 1999.

34 K. Ozbay, A. Datta, and P. Kachroo, "Modeling route choice behavior with stochastic learning automata," *Transportation Research Record: Journal of the Transportation Research Board*, vol. 1752, no. 1, pp. 38–46, 2001.

35 S. Misra and B. J. Oommen, "Dynamic algorithms for the shortest path routing problem: Learning automata-based solutions," *IEEE Transactions on Systems, Man, and Cybernetics. Part B, Cybernetics*, vol. 35, no. 6, pp. 1179–1192, 2005.

36 S. Barzegar, M. Davoudpour, M. Meybodi, A. Sadeghian, and M. Tirandazian, "Traffic signal control with adaptive fuzzy coloured petri net based on learning automata," in *Fuzzy Information Processing Society (NAFIPS), 2010 Annual Meeting of the North American*. IEEE, 2010, pp. 1–8.

37 J. A. Torkestani and M. R. Meybodi, "A learning automata-based heuristic algorithm for solving the minimum spanning tree problem in stochastic graphs," *The Journal of Supercomputing*, vol. 59, no. 2, pp. 1035–1054, 2012.

38 O. Tilak, R. Martin, and S. Mukhopadhyay, "Decentralized indirect methods for learning automata games," *IEEE Transactions on Systems, Man, and Cybernetics*, vol. 41, pp. 1213–1223, 2011.

39 W. Yuan, H. Leung, W. Cheng, and S. Chen, "Optimizing voting rule for cooperative spectrum sensing through learning automata," *IEEE Transactions on Vehicular Technology*, vol. 60, no. 7, pp. 3253–3264, 2011.

40 E. Cuevas, D. Zaldivar, and M. Pérez-Cisneros, "Seeking multi-thresholds for image segmentation with learning automata," *Machine Vision and Applications*, vol. 22, no. 5, pp. 1–11, 2011.

41 L. Jin, S. Liang, X. Luo, and M. Zhou, "Distributed and time-delayed-winner-take-all network for competitive coordination of multiple robots," *IEEE Transactions on Cybernetics*, vol. 53, no. 1, pp. 641–652, 2023.

42 P. Zhang, M. Zhou, C. Li, and A. Abusorrah, "Dynamic evolutionary game-based modeling, analysis and performance enhancement of blockchain channels," *IEEE/CAA Journal of Automatica Sinica*, vol. 10, no. 1, pp. 188–202, 2023.

43 Z. Cao, L. Zhou, C. Lin, and M. Zhou, "Solving an order batching, picker assignment, batch sequencing and picker routing problem via information integration," *Journal of Industrial Information Integration*, vol. 31, 100414, 2023, https://doi.org/10.1016/j.jii.2022.100414.

44 L. He, G. Liu, and M. Zhou, "Petri-net-based model checking for privacy-critical multiagent systems," *IEEE Transactions on Computational Social Systems*, vol. 10, no. 2, pp. 563–576, 2023.

45 Z. Zhang, H. Liu, M. Zhou, and J. Wang, "Solving dynamic traveling salesman problems with deep reinforcement learning," *IEEE Transactions on Neural Networks and Learning Systems*, vol. 34, no. 4, pp. 2119–2132, 2023.

46 L. Huang, M. Zhou, K. Hao, and H. Han, "Multirobot cooperative patrolling strategy for moving objects," *IEEE Transactions on Systems, Man, and Cybernetics: Systems*, vol. 53, no. 5, pp. 2995–3007, 2023.

7

Noisy Optimization Applications

In most real-world physical cases, the fitness or cost of a solution cannot be described by an accurate value because there is uncertainty in its measurement and/or evaluation [1].

Example 7.1 When optimizing the layout of a web-page using Simulated Annealing (SA), the fitness obtained by a solution tends to be a random variable since SA is a stochastic procedure. When training any kind of a neural network, the error rate obtained after every training run follows a statistical distribution since training a neural network is a stochastic procedure. When evolving game bots (autonomous agents), the uncertainty arises from the problem itself. In games, several factors such as the initial positions of the players and the opponent's behavior add certain stochastic components such that final scores are uncertain. The bot itself relies on probabilities to generate its behavior, and, hence, two different runs with exactly the same initial conditions and opponent tend to yield different scores.

The uncertainty of a real system can be identified as follows.

1) Changing environmental and operating conditions. Examples are the angle of attack in airfoil design, operating temperature, pressure, humidity, changing material properties and drift, etc.
2) Production tolerances and actuator imprecision. The design parameters of a product can be realized only to a certain degree of accuracy. High precision machinery is expensive, and therefore, a design less sensitive to manufacturing tolerances can reduce costs.
3) Uncertainties in the system output. These uncertainties are due to imprecision in the evaluation of the system output and the system performance. This kind

Learning Automata and Their Applications to Intelligent Systems, First Edition.
JunQi Zhang and MengChu Zhou.
© 2024 The Institute of Electrical and Electronics Engineers, Inc. Published 2024 by John Wiley & Sons, Inc.

of uncertainty includes measuring errors and all kinds of approximation errors due to the use of models instead of the real physical objects (model errors).
4) Feasibility uncertainties. The design variables must obey the uncertainties of the constraints.

Particle Swarm Optimizer (PSO) is a population-based optimization technique applied to a wide range of problems [2–9]. However, its performance deteriorates significantly in noisy environments. Some studies have addressed this issue by introducing a resampling method. Most existing methods allocate a fixed and predetermined budget of re-evaluations for every iteration, but cannot change the budget according to different environments adaptively. This chapter first introduces a PSO-LA to integrate PSO with a Learning Automaton (LA). PSO-LA utilizes LA's flexible self-adaption and automatic learning capability to learn the budget allocation for each iteration. Then, a subset scheme based-LA (subLA) embedded PSO further increases the probability of correctly finding the best particle through the pursuit on the subset of particles with better performance, yielding a method called LAPSO. LAPSO does not record the historical global best solution but finds it from the subset learned by subLA to jump out of the trapped area that may have a false global best solution. It can also adaptively consume computing budgets for every particle per iteration and, accordingly, total iterations. The version of PSO with OCBA, denoted as PSO-OCBA, is also introduced as an example of ordinal optimization applications. Through experiments on 20 large-scale benchmark functions subject to different levels of noise, LAPSO, PSO-LA, and PSO-OCBA are compared with the existing ones in both accuracy and convergence rate when they are used to solve the optimization problems in noisy environments.

7.1 Background and Motivation

Optimization is a process of searching the best solution for a problem in a reasonable time [10, 11]. It plays a major role in Computer Science, Artificial Intelligence, Operational Research, and other related fields [12]. Biologically inspired optimization algorithms, which incorporate the biological behaviors into their algorithmic framework, have thrived for multiple decades [13–21]. Various heuristic approaches have been proposed so far, such as Genetic Algorithm (GA) [22], Simulated Annealing (SA) [23, 24], Particle Swarm Optimizer (PSO) [25, 26], Ant Colony Optimization (ACO) [27–29] and Artificial Bee Colony Algorithm (ABC) [30–33].

PSO is a population-based algorithm designed by Eberhart and Kennedy in 1995 [25, 34]. It consists of a swarm of particles exploring the search space for an optimization problem and storing the best solutions found. Every particle has a position vector and a memory where the best position found is stored. The current positions of the particles change by updating velocity vectors. They are calculated according to the optimal positions found by the particles themselves and by their neighbors. As such, particles are probably guided toward the promising areas of a search space. The rather simple operation of PSO, together with other characteristics, has encouraged its adoption to tackle various optimization problems in different fields of applications [35].

However, the performance of PSO deteriorates significantly on optimization problems subject to noise [1]. In this type of problem, the true objective values of any solution are disrupted by noise. Therefore, bad solutions can be easily determined as good ones and vice versa. This characteristic particularly influences the performance of PSO adversely because particles choose the solutions as attractors in accordance with their objective values. Thus, if a bad solution disrupted by noise is selected as an attractor of the particle itself, or worse, of the entire swarm, such a solution guides the swarm toward non-optimal areas. Therefore, PSO needs to adapt the selection mechanism to take the noise into account and to vary the number of samples depending on the observed fitness difference.

Although studies [36, 37] have found PSO to be adequately stable and efficient in noisy environments, and have even suggested that noise could be beneficial for PSO, it is a clear fact that its disruptive effect needs to be explicitly tackled rather than counting on PSO to do so by [38]. A widely used approach to mitigate the effect of noise is resampling, where solutions are re-evaluated several times and their true objective values are estimated using the mean over the samples obtained. The more re-evaluations performed on a single position, the more accurate its estimated objective values. Unfortunately, accordingly, the computational cost will increase. Thus, considering a limited computational budget, the problem is then defined as how to allocate these evaluations among the solutions efficiently.

The basic resampling method is Equal Resampling (ER) [38], which re-evaluates all the particles the same number of times regardless of their qualities. As a consequence, it ends up with sacrificing the accuracy of the best particles to improve others. A more effective method [39] integrates the Optimal Computing Budget Allocation (OCBA) [40]. It first performs ER and then sequentially allocates further re-evaluations to each particle which are more likely to be the best one in the set. After that, Equal Resampling top-$\bar{N}$ (PSO-ERN) is proposed. It first uses ER to have an initial estimate of the particles and then directly allocates the extra re-evaluations of the estimated top-$\bar{N}$ particles.

PSO is integrated with Learning Automaton (PSO-LA) [41] and aims at pursuing the current best solution with less efforts in noisy environments. Unlike other mechanisms of allocating resampling re-evaluations before running, LA utilizes an action probability vector to determine particles' probabilities of being selected.

PSO-LA allows an adaptive resampling numbers in accordance with the current environment. PSO-LA is designed to identify the optimal particle and allocates most of its budgets to this one. However, due to limited computational budget, resampling methods cannot eliminate the bad effects of noise thoroughly, and thus the optimal solution found may be not as good as we expect.

To address the above issue, a subset scheme of learning automata [42] is incorporated into PSO, thereby resulting in LAPSO to further increase the probability of correctly finding the best particle through the pursuit on the a subset of particles with better performance. LAPSO does not record the historical global best solution but finds it from the subset learned by subLA to jump out of the trapped area that may have a false global best solution. In the same time, LAPSO can also adaptively consume computing budgets for every particle per iteration and, accordingly, total iteration times.

7.2 Particle Swarm Optimization

PSO as a metaheuristic inspires from the swarm theory and simulates social models such as bird flocking and fish schooling. PSO is an iterative algorithm where a swarm of particles cooperatively explore the search place sharing memories about their histories and heads toward the global best one step by step.

Particle i at iteration t consists of two $\dot{d}$-dimensional vectors that represent its velocity vector $\dot{V}_i^t = (\dot{v}_{i1}^t, \dot{v}_{i2}^t, ..., \dot{v}_{id}^t)$, and position vector $\dot{X}_i^t = (\dot{x}_{i1}^t, \dot{x}_{i2}^t, ..., \dot{x}_{id}^t)$. Apart from the two vectors, every particle has another position vector $\dot{Y}_i^t = (\dot{y}_{i1}^t, \dot{y}_{i2}^t, ..., \dot{y}_{id}^t)$, storing the best position it has ever searched. Globally, particles are partially pulled to the best position within their neighborhood. For particle i in dimension j at iteration t, its velocity and position vectors are updated as follows:

$$\dot{v}_{ij}^{t+1} = \omega \dot{v}_{ij}^t + \dot{c}_1 \dot{r}_{1j}^t \left[\dot{y}_{ij}^t - \dot{x}_{ij}^t \right] + \dot{c}_2 \dot{r}_{2j}^t \left[\dot{\hat{y}}_{ij}^t - \dot{x}_{ij}^t \right] \tag{7.1}$$

$$\dot{x}_{ij}^{t+1} = \dot{x}_{ij}^t + \dot{v}_{ij}^{t+1}, \tag{7.2}$$

where ω refers to the inertia of a particle, $\dot{c}_1$ and $\dot{c}_2$ store positive acceleration coefficients that weigh the importance of their cognitive and social components, $\dot{r}_{1j}^t$ and $\dot{r}_{2j}^t$ are random values chosen from independent uniform distributions, $\dot{y}_{ij}^t$ is the value of dimension j of the best position found by particle i, and $\dot{\hat{y}}_{ij}^t$ is the value of dimension j of the best position found by another particle within i's neighborhood $\check{N}_i$.

The existing noise-free PSO variants can be briefly classified into the following categories [9, 43]: Parameters Configurations, Topology Structures, Hybrid PSO, and Multiswarm Techniques.

7.2.1 Parameters Configurations

In [44], Shi and Eberhart introduced a new parameter called the inertia weight denoted as ω, into the original PSO. It is used to balance global and local search abilities. A linearly decreasing inertia weight is given as

$$\omega(t) = \hat{\omega} - (\hat{\omega} - \check{\omega}) \times \frac{t}{\bar{T}}. \tag{7.3}$$

where $\hat{\omega}$ and $\check{\omega}$ are the upper and lower bounds of ω, which are usually set to 0.9 and 0.4, respectively. t and $\bar{T}$ represent the current generation and max generation, respectively. In addition, the constriction factor parameter $\bar{\chi}$ [45] has been introduced into PSO for analyzing the convergence behavior. Ratnaweera et al. [46] proposed a PSO with Time-Varying Acceleration Coefficient (PSO-TVAC). The TVAC strategy is incorporated to dynamically change both $\hat{c}_1$ and $\hat{c}_2$ with time in order to achieve better control of global and local search abilities. Zhan et al. [47] proposed an Adaptive PSO (APSO) that attempts to identify the evolutionary states through the proposed Evolutionary State Estimate (ESE) module to adaptively tune $\hat{c}_1$, $\hat{c}_2$, and ω.

7.2.2 Topology Structures

A PSO's topological structure is another crucial factor to impact PSO performance. It can be static or dynamic. Several static structures have been proposed, including a ring and a von Neumann topology. Mendes et al. [48] developed a PSO variant named the Fully Informed Particle Swarm (FIPS) algorithm, in which the update of each particle is based on the positions of several neighbors. As a dynamic topology example, Liang et al. [49] proposed a Comprehensive Learning PSO (CLPSO) by suggesting that each dimension of a particle is allowed to learn from its own or others' best personal positions.

7.2.3 Hybrid PSO

Hybrid PSO [2, 50–55] combines PSO with techniques including but not limited to other evolutionary paradigms like genetic algorithm [56] and estimation of distribution[57, 58]. Gong et al. [59] propose Genetic Learning PSO (GLPSO) that uses genetic operators to generate exemplars to provide guidance to the evolution of the particles. Li et al. [60] propose Historical Memory-based PSO (HMPSO) that uses an estimation of distribution algorithm to estimate and preserve the distribution information of particles' historical promising pbest. Saxena et al. [61] propose Dynamic-PSO (DPSO) that provides chances for stagnant particles to move toward a potentially better unexplored region. Ren et al. [62] propose Scatter Learning PSO (SLPSO) that constructs an Exemplar Pool (EP) and requires particles to select their exemplars from EP using the roulette wheel rule. Chen et al.[43] propose a PSO with an Aging Leader and Challengers (ALC-PSO) in which the leader of the swarm has a growing age and a lifespan, and other individuals are allowed to challenge the leadership when a leader becomes aged.

7.2.4 Multiswarm Techniques

Classical PSO cannot provide a suitable solution for high-dimensional problems because the search space increases exponentially with the problem size. Yang et al. [63] propose a novel Segment-based Predominant Learning Swarm Optimizer (SPLSO) that divides the whole dimension into segments and lets several predominant particles guide the learning of particles. Qin et al. [64] propose an improved PSO algorithm with an Interswarm Interactive Learning PSO (IILPSO) that divides the population into two swarms and particles in a learning swarm learn from ones in the learned swarm. More multi-swarm PSO variants can be found in [65–67].

7.3 Resampling for Noisy Optimization Problems

It is common that optimization problems are influenced by noise in real world. Optimization problems in noisy environment are highly challenging for PSO [68]. A straightforward method to diminish the noisy interference is resampling, which has been a mainstream strategy for all noisy PSO variants. The causes of noise are variable, such as imprecise measurements, modeled by probability distributions or the corresponding errors [37]. These types of interference are modeled as sampling noise subjects to a Gaussian distribution [69]. Therefore, the indexes of noise are presented by its standard deviation σ and the type of noise, additive or multiplicative. Additive and multiplicative Gaussian noise may be added to function $\breve{F}(x)$:

$$\hat{F}_{+}(x) = \breve{F}(x) + N(0, \sigma^2) \tag{7.4}$$

$$\hat{F}_{\times}(x) = \breve{F}(x) \times N(1, \sigma^2). \tag{7.5}$$

Hereinafter, the true fitness value of solution $\mathbf{x}$ is represented as $\breve{F}(\mathbf{x})$, a single noisy evaluation of solution $\mathbf{x}$ is represented as $\hat{F}(\mathbf{x})$, and the estimated fitness value of solution $\mathbf{x}$ is represented as $\tilde{F}(\mathbf{x})$. Thus, in the environment where $\sigma = 0$, $\breve{F}(\mathbf{x}) = \hat{F}(\mathbf{x}) = \tilde{F}(\mathbf{x})$.

Compared with additive noise, multiplicative noise has more challenges, not only for the larger changes introduced to the objective values, but also for its deterioration exerting on the search place indirectly. In minimization problems whose values are always positive, the severity of multiplicative noise decreases as the particles improve, while in maximization ones its severity increases instead. These problems have been characterized as *backward* and *forward* ones in [70].

It is a clear fact that the introduction of resampling methods into PSO significantly mitigates the degradation caused by noise. Resampling methods can certainly reduce the misguidance from noise by re-evaluating more than once at the same position.

The essence of resampling methods is re-evaluating via different mechanisms, and then regarding the sample mean as an estimate of the true fitness value. Afterwards, PSO performs as usual, but utilizes the sample means as fitness values. However, since there is a limitation for extra re-evaluations budget, it is necessary to introduce an efficient resampling method into PSO hunting for a tradeoff between the number of re-evaluations and iterations. This chapter reviews four resampling methods, i.e., Equal Resampling, Optimal Computing Budget Allocation, Equal Resampling top-$\bar{N}$ particles, and Learning Automaton.

ER is the most basic resampling method, which allocates all particles with the equal initial re-evaluations $\hat{B}_0$ and extra re-evaluations $\hat{B}_\Delta$ regardless of their solution qualities. In this case, the accuracy of the whole swarm gets improved, but it sacrifices some re-evaluations put on worse particles.

ERN [38] is a simpler resampling method. It performs ER first and then allocates the extra budgets equally to $\bar{N}$ particles which have better performance. In PSO-ERN, particles update their personal best solutions if the current ones are either better or more accurate. Besides, the best global solution is selected only among the most accurate ones.

7.4 PSO-Based LA and OCBA

OCBA is proposed in the field of simulation optimization to enhance the efficiency of ordinal optimization for discrete-event simulation designs [40] and introduced in Chapter 5. The designs refer to as those particles whose performances are disrupted by noisy environments and call for re-evaluation to obtain a more accurate estimating values. Despite various environments, OCBA is unable to change the budget in accordance with the environment adaptively at every iteration. It evaluates the mean and variances of the particles via ER on each particle first. Next, it allocates an extra budget $\hat{B}_\Delta$ sequentially in groups of $\hat{B}_\Omega$ evaluations. The selection is based on their means and variances, both of which are updated after each evaluation. Finally, the approximate probability of correct selection is asymptotically maximized when $\hat{B}_\Delta$ is used according to the following equations:

$$\frac{\mathbb{N}_i(t)}{\mathbb{N}_j(t)} = \left(\frac{\hat{\sigma}_i(t)/(\hat{\delta}_{i,m}(t))}{\hat{\sigma}_j(t)/(\hat{\delta}_{j,m}(t))} \right)^2, \ \forall i, j \in \mathbb{Z}_{\bar{N}} \text{ and } i \neq j \neq m, \tag{7.6}$$

$$\mathbb{N}_m(t) = \hat{\sigma}_m(t) \sqrt{\sum_{i=1, i\neq m}^{\bar{N}} \frac{\mathbb{N}_i^2(t)}{(\hat{\sigma}_i(t))^2}}, \tag{7.7}$$

where $\mathbb{N}_i(t)$ refers to the total number of evaluations allocated to particles i. $\hat{d}_i(t)$ and $\hat{\sigma}_i(t)$ store its sample mean and variance, respectively, and m stands for the best solution (selected as the one with the lowest mean). $\hat{\delta}_{i,m}(t) = \hat{d}_i(t) - \hat{d}_m(t)$. The main purpose of OCBA is to decide how to allocate limited computational budget $\hat{B}_A$ among the set of designs at every iteration. For further information, a reader is encouraged to read [69–71].

The PSO-LA [41] utilizes an action probability vector of learning automata to determine the probability for a particle to select. At the beginning of extra re-evaluations, LA allocates different particles with different probabilities according to their performance after $\hat{B}_0$ times initial re-evaluations. The probability vector $\mathbf{P} = (p_1, p_2, ..., p_{\bar{N}})$ is designed according to (7.8) and (7.9), μ_i represents the average fitness value after $\hat{B}_0$ times initial re-evaluations. LA always rewards the best particle by increasing its probability additional and gives penalties to others, i.e. decreasing the action probabilities. Either the best particle's probability reaches the threshold or the number of resampling times arrives at the upper limit, LA stops and PSO continues its next iteration.

$$\omega_i = \frac{\hat{\mu} - \mu_i}{\hat{\mu} - \check{\mu}} \tag{7.8}$$

$$p_i = \frac{\omega_i}{\sum_{j=1}^{\bar{N}} \omega_j}. \tag{7.9}$$

A new class of learning automata for selecting an optimal subset [42] in noisy environment can be used to further help the particle swarm to resist noise. Different from the traditional aim of LA, this subset scheme selects an optimal subset of all actions, rather than a single optimal action. Its objective is to identify the optimal subset: the top-$\bar{N}$ out of $\tilde{N}$ particles. Based on traditional continuous pursuit and discretized pursuit learning schemes, there are four subset-pursuit learning schemes for selecting the optimal subset, called continuous equal pursuit, discretized equal pursuit, continuous unequal pursuit, and discretized unequal pursuit learning schemes, respectively. Studies have proved that discretized equal pursuit algorithm, which is used in this chapter, outperforms others [42]. The updating rule of this scheme is (7.10) and (7.11), where $\check{S}_m$ stores the currently estimated subset. The subset scheme is used to select a subset of promising particles in PSO.

$$p_i(t+1) = max(0, p_i(t) - \Delta), \forall i \notin \check{S}_m \tag{7.10}$$

$$p_i(t+1) = \frac{1}{|\check{S}_m|} \left(1 - \sum_{j \notin \check{S}_m} p_j(t+1) \right), \forall i \in \check{S}_m. \tag{7.11}$$

Because of limited computing budgets, the DP_{RP} version subLA by using the discretized equal pursuit scheme is integrated with PSO, in order to reach the threshold as soon as possible. A subset stores the current Top-$\tilde{N}$ particles and is updated after each re-evaluation.

Example 7.2 LAPSO for noisy optimization: At first, each particle receives only one initial evaluation. After that, PSO uses the estimated fitness μ_i as a function value. Every particle gets an equal probability $p_i = \frac{1}{\tilde{N}}$ ($i = 1, 2, ..., \tilde{N}$) at the beginning of each iteration from subLA. For each sampling, the response is 1, if the estimated fitness of the selected particle is the best one in the subset. subLA would always reward the solutions in its subset, while others receive a penalty Δ over probabilities correspondingly. This select-and-discard mechanism allows subLA to modify the subset in accordance with particles' quality in a timely fashion. As a consequence, LAPSO allocates more budgets to promising ones, rather than only a single one. If (i) the sum of probabilities of particles in a subset reaches the threshold or (ii) the number of resampling times arrives at the upper limit, subLA stops and the PSO continues to the next iteration. Before continuing to its next iteration, the global best solution is selected from the final subset of the current iteration compulsorily instead of the historical global best one. Such way helps the swarm jump out of the trapped area that may have a false global best solution [72]. Finally, in some iterations, subLA may reach the threshold before using all budgets up. This is because subLA conducts fewer re-evaluations in simple environments than in complex ones adaptively. Therefore, globally, these budgets could be utilized for more explorations. In addition, in case of looping in subLA, it has a maximum number of re-evaluation times per iteration.

Specifically, the trapped area that may have a false global best solution can mislead the whole swarm from moving toward a better position because of a slow-changed or unchanged global best solution. Therefore, the compulsorily updated global best solution proposed in this chapter is based on the fact that the whole swarm tends to be misled due to inaccurate historical information in a noisy environment.

Compared with other resampling methods, PSO-OCBA cannot change its allocation according to the current environment. PSO-ERN allocates all its extra budgets to those particles that perform better at the first five re-evaluations, which lacks objectivity and accuracy. PSO-LA allocates too many budgets to a single particle, which might not be as good as expected. In this circumstance, it is reasonable to allocate some re-evaluations to estimated-suboptimal particles, hoping them for a better performance later. Thus, LAPSO is preferable based on the above analysis.

LAPSO ALGORITHM

Input: Number of particles $\tilde{N}$, Positive acceleration coefficients $\acute{c}_1, \acute{c}_2$, Inertia weight ω, number of allowable actions r, resolution parameter n, amount of initial samples $\hat{B}_0$, additional budget $\hat{B}_A$, maximum number of re-evaluation times per iteration $\hat{B}_A$.

Output: Best position found by particles.

Initialize particle swarm;

Update the best particle;

subLA probability vector; subLA counter, $\hat{M} \leftarrow 0$.

while (stopping criterion not met) **do**

    for (i = 1, ... , $\tilde{N}$) **do**

        Evaluate $\dot{X}_i$, get estimated fitness value $\tilde{F}(\dot{X}_i^t)$

        Initial subLA with $p_i = \frac{1}{p}$

    end for

    Call algorithm subLA for $\tilde{F}(\dot{X}_i^t)$ (i = 1, ... , $\tilde{N}$)

    while $\hat{M} < \hat{B}_A$ or sum(p_{subset}) < threshold **do**

        Choose one particle n,

        Evaluate $\tilde{F}(\dot{X}_n^t)$ based on the initial replications

        give a penalty on probabilities to particles which are not selected in subset using (10)

        reward subset $\check{S}_j$ using (11)

    end while

    for (i = 1, ... , $\tilde{N}$) **do**

        if $\tilde{F}(\dot{X}_i^t) < \tilde{F}(\dot{Y}_i^{t-1})$ **then**

            $\dot{Y}_i^t \leftarrow \dot{X}_i^t$

        else

            $\dot{Y}_i^t \leftarrow \dot{Y}_i^{t-1}$

        end if

    end for

    for (j = 1, ... , $\tilde{N}$) **do**

        $\hat{Y}_i^t \leftarrow \arg\min \check{F}(\check{S}_i^t)$ $_{(\dot{Y}_i^t \in \mathbb{Z}_i)}$

    end for

    for (i = 1, ... , $\tilde{N}$) **do**

        Update $\dot{V}_i^{t+1}$ and $\dot{X}_i^{t+1}$ using (1) and (2)

    end for

end while

END

7.5 Simulations Studies

The benchmark functions upon which we appraise the performance of the algorithms are those from the CEC'2010 Special Session and Competition on Large-Scale Global Optimization [73]. It offers 20 functions whose fitness values are all positive and the global minimum is $\check{F}(x) = 0$. These benchmarks consist of five classical functions, named Elliptic ($F_{01}, F_{04}, F_{09}, F_{14}$), Rastrigin ($F_{02}, F_{05}, F_{10}, F_{15}$), Ackley ($F_{03}, F_{06}, F_{11}, F_{16}$), Schwefel ($F_{19}, F_{07}, F_{12}, F_{17}$), and Rosenbrock ($F_{20}, F_{08}, F_{13}, F_{18}$).

For comparison, PSO, PSO-ER, PSO-OCBA, PSO-ERN, PSO-LA, and LAPSO share the same experimental vectors as those in [38]. The list of parameter values is shown in Table 7.1. The experiments are implemented on these benchmark functions subject to different levels of multiplicative Gaussian noise decided by $\sigma \in \{0.06, 0.12, 0.18, 0.24, 0.30\}$, but the total number of noise samples is limited to stay within 3σ in order to ensure positive fitness values. The algorithms to be evaluated upon these benchmarks are PSO, PSO-ER, PSO-OCBA, PSO-ERN, PSO-LA, and LAPSO, each of which performs 50 times of independent runs on every problem subject to different levels of noise. All the algorithms have the same initial conditions of each run, except for the initial re-evaluation times for LAPSO.

Specifically, although particles in subLA receive little budget at first, they still have the probability to get re-evaluated. Even if the initial performances of some promising particles are bad, their performances could be updated during

Table 7.1 Parameter values.

Parameter	Values
Independent runs	15 000 function evaluations
Number of particles	50 in R^{1000} with star topology
Acceleration&Inertia	$\hat{c}_1 = \hat{c}_2 = 1.49618, \omega = 0.729844$
Maximum velocity	$0.25 \cdot |\dot{x}_{max} - \dot{x}_{min}|$
Severity of noise	$\sigma = \{0.06, 0.12, 0.18, 0.24, 0.30\}$
Computational budgets	$\hat{B}_0 = 5, \hat{B}_\Delta = 50, \hat{B}_\Lambda = 250$
PSO-ER	$\vec{b}_i = \hat{B}_\Delta/50 = 1$
PSO-OCBA	$\hat{B}_\Omega = 5$
PSO-ERN	$\bar{N} = 2, \vec{b}_i = \hat{B}_\Delta/\bar{N}$
PSO-LA	Threshold $= 0.7$
LAPSO	Threshold $= 0.9$ Subset capacity $= 3$ $\hat{B}_0 = 1$

the automatic learning process of subLA gradually. As for other resampling methods, such as OCBA, its allocation is also based on probability, but particles' probabilities are decided by the variance and mean of the initial re-evaluations fixedly, so if the initial re-evaluations are not accurate enough, much budget would be allocated to unpromising particles. As a distinct character that OCBA lacks subLA searches for the optimal parameter vector in the space of probability distributions over the parameter space, rather than in the parameter space itself [74], such as OCBA.

The star topology and acceleration as well as inertia coefficients mentioned in [38] are used in PSO. Each swarm has a total computational budget of 15 000 function evaluations which results in $15\,000/50 = 300$ iterations in the case of PSO (50 is the number of particles). OCBA divides such a budget into $\hat{B}_0 = 5$ evaluations for each particle (as suggested in [71]) and $\hat{B}_A = 50$ re-evaluations (equivalent to one iteration) allocated via a different mechanism. Thus, common integrations with resampling can execute $\frac{15\,000}{(5 \times 50 + 50)} = 50$ iterations only, but due to the re-evaluation, the fitness values are more accurate. Specifically, the accuracy of the solutions in PSO-ER is based on 6 samples only, while under the same circumstance those in LAPSO are based on $\frac{\text{Threshold Probability} - \bar{N} \times \frac{1}{\bar{N}}}{A \times (\hat{N} - \bar{N})}$ samples at least.

In order to illustrate the most efficient capacity for a subset, we conduct experiments on LAPSO with different capacities of $\{1, 2, 3, 4, 5, 10\}$ and $\sigma = 0.30$. The results of five typical functions are shown in Figs 7.1–7.5. According to the result,

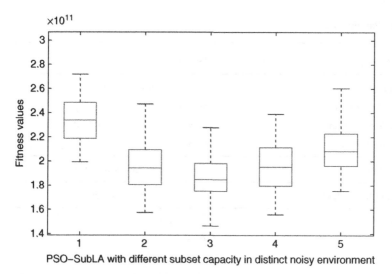

PSO–SubLA with different subset capacity in distinct noisy environment

Figure 7.1 The comparison of fitness values (left axis) of different subset capacity (bottom axis) under $F_{09}(Elliptic)$.

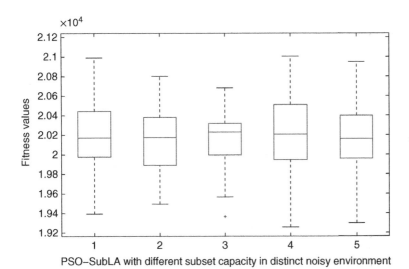

Figure 7.2 The comparison of fitness values (left axis) of different subset capacity (bottom axis) under $F_{10}(Rastrigin)$.

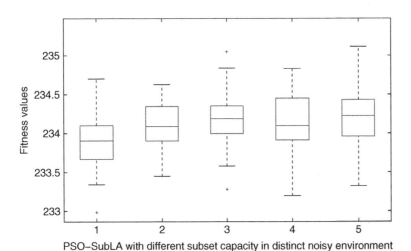

Figure 7.3 The comparison of fitness values (left axis) of different subset capacity (bottom axis) under $F_{11}(Ackley)$.

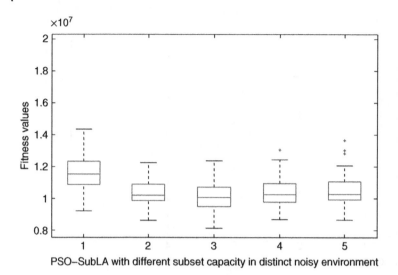

Figure 7.4 The comparison of fitness values (left axis) of different subset capacity (bottom axis) under $F_{12}(Schwefel)$.

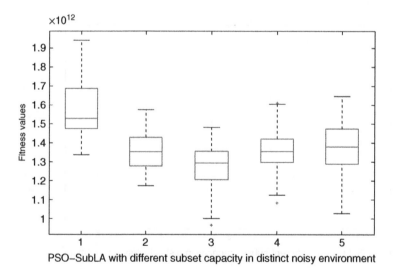

Figure 7.5 The comparison of fitness values (left axis) of different subset capacity (bottom axis) under $F_{13}(Rosenbrock)$.

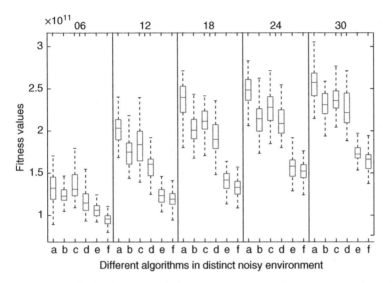

Figure 7.6 The comparison of fitness values (left axis) of PSO, PSO-ER, PSO-OCBA, PSO-LA, PSO-ERN, and LAPSO (bottom axis) under F_1.

when the subset capacity equals 3, the result achieves the best, which is the same statistic as recommended in [42].

The results from the experiments of the state-of-the-art methods and the proposed one are shown in Figs 7.6–7.25. According to the results, it is necessary to introduce resampling methods to improve the performance of PSO in noisy environments.

Globally, PSO-LA performs 19 out of 20 benchmark functions better than PSO, PSO-ER, PSO-OCBA, PSO-ERN and PSO-LA. In terms of Rastrigin, Ackley, Schwefel, and Rosenbrock class, subLA shows a better performance in all of their functions, while as for Elliptic class, subLA is inferior to PSO-ERN on F_{04} only.

Notice that all algorithms further deteriorate as the level of noise increases. This can be seen by each boxplot across the sections that indicate the level of noise at the top axis. The "abcdef", which are put at the bottom axis, stand for PSO, PSO-ER, PSO-OCBA, PSO-LA PSO-ERN, and LAPSO, respectively. The left axis stores the classification of functions and the real fitness values found by different algorithms. As for these benchmark functions, the values are the smaller, the better.

According to the experiments under 15 000 evaluations, LA's and subLA's iteration counts usually lie in the range from 52 to 58, while others perform fixedly 50 iterations. PSO-LA and LAPSO thus allow a tradeoff between exploration and exploitation. They spend in the range of $(\frac{52 \times 50}{15\,000} = 17.33\%, \frac{58 \times 50}{15\,000} = 19.33\%)$ on exploration, which is based on different environments adaptively, while OCBA, ER,

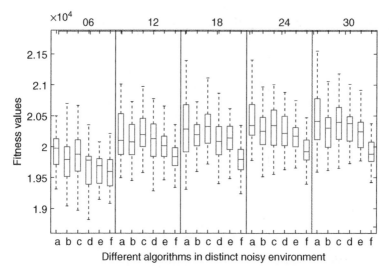

Figure 7.7 The comparison of fitness values (left axis) of PSO, PSO-ER, PSO-OCBA, PSO-LA, PSO-ERN, and LAPSO (bottom axis) under F_2.

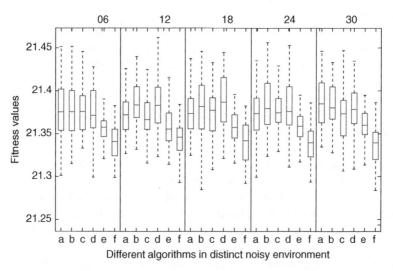

Figure 7.8 The comparison of fitness values (left axis) of PSO, PSO-ER, PSO-OCBA, PSO-LA, PSO-ERN, and LAPSO (bottom axis) under F_3.

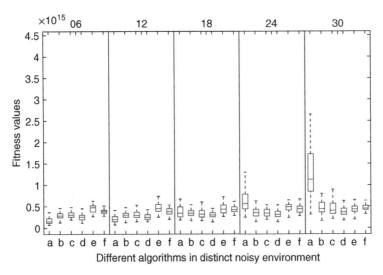

Figure 7.9 The comparison of fitness values (left axis) of PSO, PSO-ER, PSO-OCBA, PSO-LA, PSO-ERN, and LAPSO (bottom axis) under F_4.

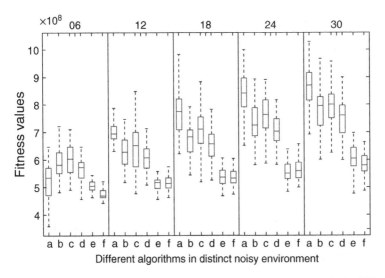

Figure 7.10 The comparison of fitness values (left axis) of PSO, PSO-ER, PSO-OCBA, PSO-LA, PSO-ERN, and LAPSO (bottom axis) under F_5.

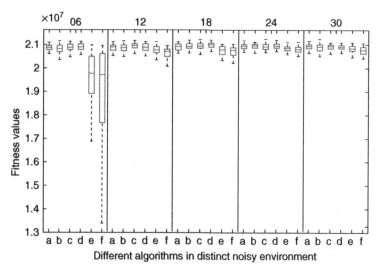

Figure 7.11 The comparison of fitness values (left axis) of PSO, PSO-ER, PSO-OCBA, PSO-LA, PSO-ERN, and LAPSO (bottom axis) under F_6.

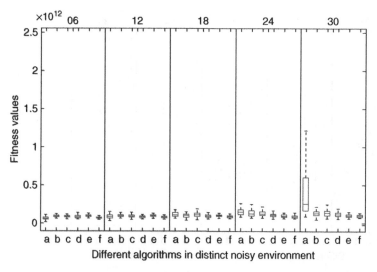

Figure 7.12 The comparison of fitness values (left axis) of PSO, PSO-ER, PSO-OCBA, PSO-LA, PSO-ERN, and LAPSO (bottom axis) under F_7.

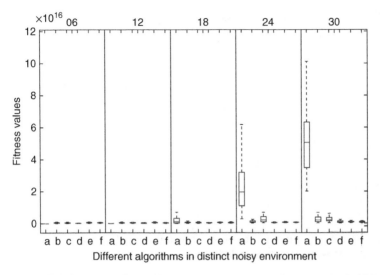

Figure 7.13 The comparison of fitness values (left axis) of PSO, PSO-ER, PSO-OCBA, PSO-LA, PSO-ERN, and LAPSO (bottom axis) under F_8.

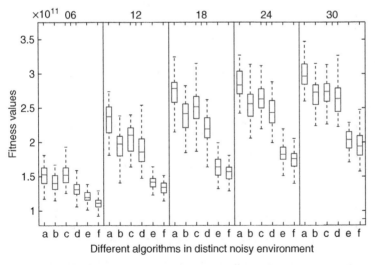

Figure 7.14 The comparison of fitness values (left axis) of PSO, PSO-ER, PSO-OCBA, PSO-LA, PSO-ERN, and LAPSO (bottom axis) under F_9.

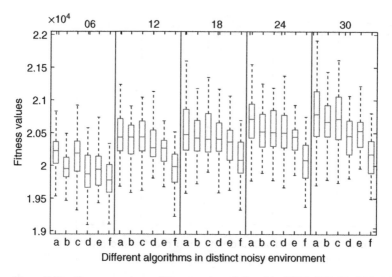

Figure 7.15 The comparison of fitness values (left axis) of PSO, PSO-ER, PSO-OCBA, PSO-LA, PSO-ERN, and LAPSO (bottom axis) under F_{10}.

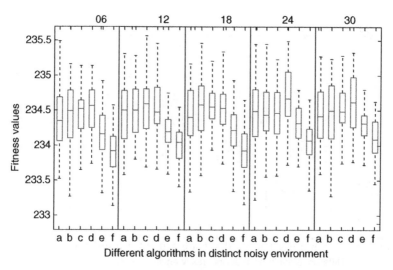

Figure 7.16 The comparison of fitness values (left axis) of PSO, PSO-ER, PSO-OCBA, PSO-LA, PSO-ERN, and LAPSO (bottom axis) under F_{11}.

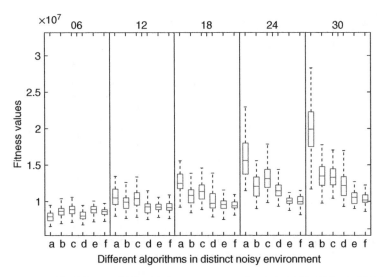

Figure 7.17 The comparison of fitness values (left axis) of PSO, PSO-ER, PSO-OCBA, PSO-LA, PSO-ERN, and LAPSO (bottom axis) under F_{12}.

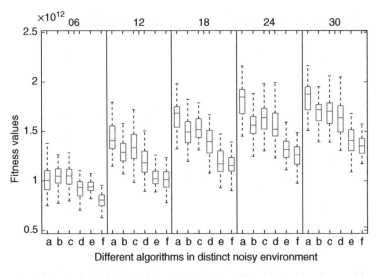

Figure 7.18 The comparison of fitness values (left axis) of PSO, PSO-ER, PSO-OCBA, PSO-LA, PSO-ERN, and LAPSO (bottom axis) under F_{13}.

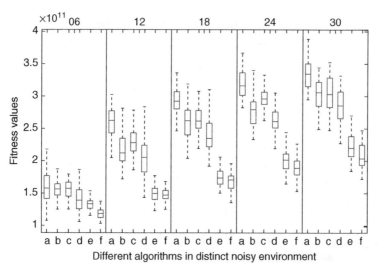

Figure 7.19 The comparison of fitness values (left axis) of PSO, PSO-ER, PSO-OCBA, PSO-LA, PSO-ERN, and LAPSO (bottom axis) under F_{14}.

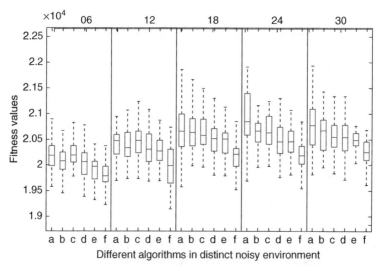

Figure 7.20 The comparison of fitness values (left axis) of PSO, PSO-ER, PSO-OCBA, PSO-LA, PSO-ERN, and LAPSO (bottom axis) under F_{15}.

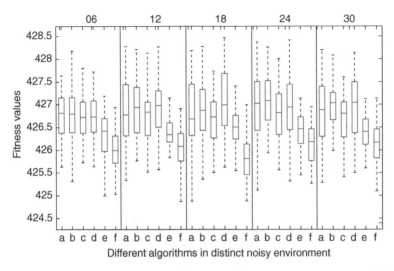

Figure 7.21 The comparison of fitness values (left axis) of PSO, PSO-ER, PSO-OCBA, PSO-LA, PSO-ERN, and LAPSO (bottom axis) under F_{16}.

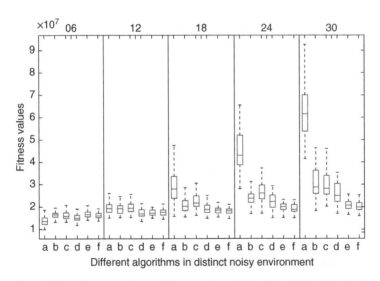

Figure 7.22 The comparison of fitness values (left axis) of PSO, PSO-ER, PSO-OCBA, PSO-LA, PSO-ERN, and LAPSO (bottom axis) under F_{17}.

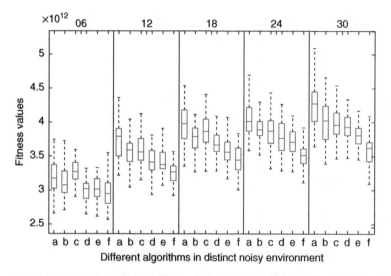

Figure 7.23 The comparison of fitness values (left axis) of PSO, PSO-ER, PSO-OCBA, PSO-LA, PSO-ERN, and LAPSO (bottom axis) under F_{18}.

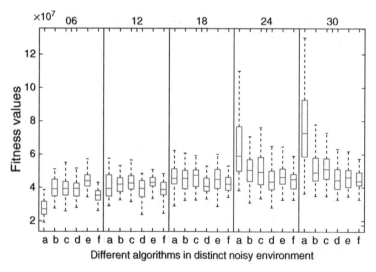

Figure 7.24 The comparison of fitness values (left axis) of PSO, PSO-ER, PSO-OCBA, PSO-LA, PSO-ERN, and LAPSO (bottom axis) under F_{19}.

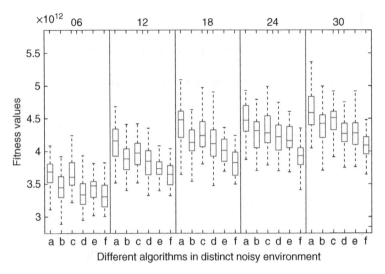

Figure 7.25 The comparison of fitness values (left axis) of PSO, PSO-ER, PSO-OCBA, PSO-LA, PSO-ERN, and LAPSO (bottom axis) under F_{20}.

and ERN spend only $\frac{50 \times 50}{15\,000} = 16.67\%$ of its evaluation budgets for this part. In this case, LA and subLA would make more global search possible while guaranteeing the quality of exploitation.

Besides, the threshold of LA is 0.7, while subLA's is 0.9. The threshold of LA measures the highest probability among all particles, but the threshold of subLA is the sum of probabilities in a subset. Obviously, the sum of probabilities tends to be easier to achieve the threshold than a single probability. In order to ensure adequate budgets for a subset, we thus use a bigger threshold than LA's.

7.6 Summary

This chapter presents learning automata and ordinal optimization-based PSOs to solve optimization problems subject to noisy fitness evaluation. The hybrid approach integrates PSO with a LA that uses a subset scheme to allocate computing budgets in noisy environments adaptively and efficiently. Overall, compared with its state-of-the-art PSO peers, the introduced methods save more computing budgets than other resampling methods, while ensuring the accuracy of result at the same time. Ranking and selection are the key to intelligent optimization, especially in stochastic environments. The learning automata and ordinal optimization own their innate advantages to rank and select optimal candidates in stochastic environments. They are well illustrated through our benchmark studies. The promising work of applying learning automata and ordinal optimization to other population-based optimizers [75–79] remain to be performed.

7.7 Exercises

1 How does the traditional PSO solve noisy optimization problems?

2 How to combine LA with PSO and what effects such combination can achieve?

3 What is the difference between PSO-LA and LAPSO?

4 Describe the algorithmic process of LAPSO in detail.

5 Think about how to further improve the anti-noise ability of PSO.

6 Run LAPSO with different parameters for different optimization problems to test the sensitivity of its performance vs. the parameter settings.

7 Explain the advantage of LA used to improve PSO.

8 Design an algorithm incorporating OCBA into PSO to handle the noise issue in fitness evaluation.

9 What is the difference between using OCBA and LA to enhance the PSO's performance to noise?

10 Incorporate LA_{OCBA} into PSO to solve a noisy optimization problem and compare its performance with PSO-LA and LAPSO.

11 Design an algorithm that incorporates OCBA/LA into another one, e.g., ABC, Genetic Algorithm, and differential evolution, to solve a noisy optimization problem.

References

1 C. Schmidt, J. Branke, and S. E. Chick, "Integrating techniques from statistical ranking into evolutionary algorithms," in Applications of Evolutionary Computing. Springer, 2006, pp. 752–763.

2 J. Bi, H. Yuan, S. Duanmu, M. Zhou, and A. Abusorrah, "Energy-optimized partial computation offloading in mobile-edge computing with genetic simulated-annealing-based particle swarm optimization," *IEEE Internet of Things Journal*, vol. 8, no. 5, pp. 3774–3785, 2021.

3 M. Ghahramani, A. O'Hagan, M. Zhou, and J. Sweeney, "Intelligent geodemographic clustering based on neural network and particle swarm optimization," *IEEE Transactions on Systems, Man, and Cybernetics: Systems*, vol. 52, no. 6, pp. 3746–3756, 2022.

4 L. Li and M. Zhou, "Sustainable manufacturing systems: An energy perspective," John Wiley & Sons, 2022.

5 Y. Wang and X. Zuo, "An effective cloud workflow scheduling approach combining PSO and idle time slot-aware rules," *IEEE/CAA Journal of Automatica Sinica*, vol. 8, no. 5, pp. 1079–1094, 2021.

6 J. Tang, G. Liu, and Q. Pan, "A review on representative swarm intelligence algorithms for solving optimization problems: Applications and trends," *IEEE/CAA Journal of Automatica Sinica*, vol. 8, no. 10, pp. 1627–1643, 2021.

7 J. Zhang, Y. Lu, L. Che, and M. Zhou, "Moving-distance-minimized PSO for mobile robot swarm," *IEEE Transactions on Cybernetics*, vol. 52, no. 9, pp. 9871–9881, 2021.

8 X. Xu, J. Li, M. Zhou, J. Xu, and J. Cao, "Accelerated two-stage particle swarm optimization for clustering not-well-separated data," *IEEE Transactions on Systems, Man, and Cybernetics: Systems*, vol. 50, no. 11, pp. 4212–4223, 2020.

9 Y. Cao, H. Zhang, W. Li, M. Zhou, Y. Zhang, and W. A. Chaovalitwongse, "Comprehensive learning particle swarm optimization algorithm with local search for multimodal functions," *IEEE Transactions on Evolutionary Computation*, vol. 23, no. 4, pp. 718–731, 2019.

10 A. D. Belegundu and T. R. Chandrupatla, "Optimization concepts and applications in engineering," Cambridge University Press, 2019.

11 M. Zhou, H.-X. Li, and M. Weijnen, "Contemporary issues in systems science and engineering," John Wiley & Sons, 2015.

12 K. Deb, "Optimization for engineering design: Algorithms and examples," cose.math.bas.bg, 1996.

13 W. Dong and M. Zhou, "A supervised learning and control method to improve particle swarm optimization algorithms," *IEEE Transactions on Systems, Man, and Cybernetics: Systems*, vol. 47, no. 7, pp. 1135–1148, 2017.

14 W. Gu, Y. Yu, and W. Hu, "Artificial bee colony algorithmbased parameter estimation of fractional-order chaotic system with time delay," *IEEE/CAA Journal of Automatica Sinica*, vol. 4, no. 1, pp. 107–113, 2017.

15 W. Han, J. Xu, M. Zhou, G. Tian, P. Wang, X. Shen, and E. Hou, "Cuckoo search and particle filter-based inversing approach to estimating defects via magnetic flux leakage signals," *IEEE Transactions on Magnetics*, vol. 52, no. 4, pp. 1–11, 2016.

16 Q. Kang, J. Wang, M. Zhou, and A. Ammari, "Centralized charging scheduling for electric vehicles with battery-swapping on spot pricing via heuristic algorithms," *IEEE Transactions on Intelligent Transportation Systems*, vol. 17, no. 3, pp. 659–669, 2016.

17 K. Xing, L. Han, and M. Zhou, "Deadlock-free genetic scheduling algorithm for automated manufacturing systems based on deadlock control policy," *IEEE Transactions on Systems, Man, and Cybernetics, Part B (Cybernetics)*, vol. 42, no. 3, p. 603–615, 2012.

18 Z. J. Ding, J. J. Liu, Y. Q. Sun, C. J. Jiang, and M. Zhou, "A transaction and QoS-aware service selection approach based on genetic algorithm," *IEEE Transactions on Systems, Man, and Cybernetics: Systems*, vol. 45, no. 7, pp. 1035–1046, 2015.

19 H. Y. J. Bi, J. Zhai, M. Zhou, and H. V. Poor, "Self-adaptive bat algorithm with genetic operations," *IEEE/CAA Journal of Automatica Sinica*, vol. 9, no. 7, pp. 1284–1294, 2022.

20 X. Zuo, C. Chen, W. Tan, and M. Zhou, "Vehicle scheduling of an urban bus line via an improved multiobjective genetic algorithm," *IEEE Transactions on Intelligent Transportation Systems*, vol. 16, no. 2, pp. 1030–1041, 2015.

21 Y. Hou, N. Wu, M. Zhou, and Z. Li, "Pareto-optimization for scheduling of crude oil operations in refinery via genetic algorithm," *IEEE Transactions on Systems, Man, and Cybernetics: Systems*, vol. 47, no. 3, pp. 517–530, 2017.

22 K.-S. Tang, K. Man, S. Kwong, and Q. He, "Genetic algorithms and their applications," *IEEE Signal Processing Magazine*, vol. 13, no. 6, pp. 22–37, 1996.

23 S. Kirkpatrick, "Optimization by simulated annealing: Quantitative studies," *Journal of Statistical Physics*, vol. 34, no. 5-6, pp. 975–986, 1984.

24 Y. Zhou, W. Xu, Z.-H. Fu, and M. Zhou, "Multi-neighborhood simulated annealing-based iterated local search for colored traveling salesman problems," *IEEE Transactions on Intelligent Transportation Systems*, vol. 23, no. 9, pp. 16072–16082, 2022.

25 J. Kennedy and R. Eberhart, "Particle swarm optimization," in *Proceedings of the IEEE International Conference on Neural Networks*, pp. 1942–1948, 1995.

26 J. Chen, X. Luo, and M. Zhou, "Hierarchical particle swarm optimization-incorporated latent factor analysis for large-scale incomplete matrices," *IEEE Transactions on Big Data*, vol. 8, no. 6, pp. 1524–1536, 2022.

27 M. Dorigo, V. Maniezzo, and A. Colorni, "Ant system: Optimization by a colony of cooperating agents," *IEEE Transactions on Systems, Man, and Cybernetics, Part B (Cybernetics)*, vol. 26, no. 1, pp. 29–41, 1996.

28 Z. Wang, S. Gao, M. Zhou, S. Sato, J. Cheng, and J. Wang, "Information-theory-based nondominated sorting ant colony optimization for multiobjective feature selection in classification," *IEEE Transactions on Cybernetics*, doi: 10.1109/TCYB.2022.3185554, 2022.

29 Y. Feng, M. Zhou, G. Tian, Z. Li, Z. Zhang, Q. Zhang, and J. Tan, "Target disassembly sequencing and scheme evaluation for CNC machine tools using improved multiobjective ant colony algorithm and fuzzy integral," *IEEE Transactions on Systems, Man, and Cybernetics: Systems*, vol. 49, no. 12, pp. 2438–2451, 2019.

30 D. Karaboga and B. Basturk, "A powerful and efficient algorithm for numerical function optimization: Artificial bee colony (ABC) algorithm," *Journal of Global Optimization*, vol. 39, no. 3, pp. 459–471, 2007.

31 G. Tian, Y. Ren, Y. Feng, M. C. Zhou, H. Zhang, and J. Tan, "Modeling and planning for dual-objective selective disassembly using and/or graph and discrete artificial bee colony," *IEEE Transactions on Industrial Informatics*, vol. 15, no. 4, pp. 2456–2468, 2019.

32 H. Jia, H. Miao, G. Tian, M. Zhou, Y. Feng, Z. Li, and J. Li, "Multiobjective bike repositioning in bike-sharing systems via a modified artificial bee colony algorithm," *IEEE Transactions on Automation Science and Engineering*, vol. 17, no. 2, pp. 909–920, 2020.

33 H. Liu, M. Zhou, X. Guo, Z. Zhang, N. Bin, and T. Tang, "Timetable optimization for regenerative energy utilization in subway systems," *IEEE Transactions on Intelligent Transportation Systems*, vol. 20, no. 9, pp. 3247–3257, 2019.

34 R. Eberhart and J. Kennedy, "A new optimizer using particle swarm theory," in Proceedings of the 6th International Symposium on Micro Machine and Human Science, Nagoya, Japan, 1995, pp. 39–43.

35 R. Poli, "Analysis of the publications on the applications of particle swarm optimisation," *Journal of Artificial Evolution and Applications*, vol. 2008, p. 3, 2008.

36 K. Parsopoulos and M. Vrahatis, "Particle swarm optimizer in noisy and continuously changing environment," *Methods*, vol. 5, no. 6, p. 23, 2001.

37 K. Parsopoulos and M. Vrahatis, "Particle swarm optimization for imprecise problems," in Scattering and Biomedical Engineering, Modeling and Applications, pp. 254–264, 2002.

38 J. Rada-Vilela, M. Zhang, and M. Johnston, "Resampling in particle swarm optimization," in *2013 IEEE Congress on Evolutionary Computation*, 2013, pp. 947–954.

39 J. Rada-Vilela, M. Zhang, and M. Johnston, "Optimal computing budget allocation in particle swarm optimization," in *Proceeding of the Fifteenth Annual Conference on Genetic and Evolutionary Computation Conference*. ACM, 2013, pp. 81–88.

40 C.-H. Chen, J. Lin, E. Yücesan, and S. E. Chick, "Simulation budget allocation for further enhancing the efficiency of ordinal optimization," *Discrete Event Dynamic Systems: Theory and Applications*, vol. 10, pp. 251–270, 2000.

41 J. Zhang, L. Xu, J. Li, Q. Kang, and M. Zhou, "Integrating particle swarm optimization with learning automata to solve optimization problems in noisy environment," in *2014 IEEE International Conference on Systems, Man and Cybernetics (SMC)*, 2014, pp. 1432–1437.

42 J. Zhang, Z. Li, Q. Kang, and M. Zhou, "A new class of learning automata for selecting an optimal subset," in *2014 IEEE International Conference on Systems, Man and Cybernetics (SMC).* IEEE, 2014, pp. 3429–3434.

43 W.-N. Chen, J. Zhang, Y. Lin, N. Chen, Z.-H. Zhan, H. S.-H. Chung, Y. Li, and Y.-h. Shi, "Particle swarm optimization with an aging leader and challengers." *IEEE Transactions on Evolutionary Computation*, vol. 17, no. 2, pp. 241–258, 2013.

44 Y. Shi and R. Eberhart, "A modified particle swarm optimizer," in *Proceedings of IEEE World Congress on Computational Intelligence.* IEEE, 1998, pp. 69–73.

45 M. Clerc and J. Kennedy, "The particle swarm-explosion, stability, and convergence in a multidimensional complex space," *IEEE Transactions on Evolutionary Computation*, vol. 6, no. 1, pp. 58–73, 2002.

46 A. Ratnaweera, S. Halgamuge, and H. C. Watson, "Self-organizing hierarchical particle swarm optimizer with time-varying acceleration coefficients," *IEEE Transactions on Evolutionary Computation*, vol. 8, no. 3, pp. 240–255, 2004.

47 Z.-H. Zhan, J. Zhang, Y. Li, and H.-H. Chung, "Adaptive particle swarm optimization," *IEEE Transactions on Systems, Man, and Cybernetics, Part B (Cybernetics)*, vol. 39, no. 6, pp. 1362–1381, 2009.

48 R. Mendes, J. Kennedy, and J. Neves, "The fully informed particle swarm: Simpler, maybe better," *IEEE Transactions on Evolutionary Computation*, vol. 8, no. 3, pp. 204–210, 2004.

49 J. J. Liang, A. K. Qin, P. N. Suganthan, and S. Baskar, "Comprehensive learning particle swarm optimizer for global optimization of multimodal functions," *IEEE Transactions on Evolutionary Computation*, vol. 10, no. 3, pp. 281–295, 2006.

50 G. Tian, Y. Ren, and M. Zhou, "Dual-objective scheduling of rescue vehicles to distinguish forest fires via differential evolution and particle swarm optimization combined algorithm," *IEEE Transactions on Intelligent Transportation Systems*, vol. 17, no. 11, pp. 3009–3021, 2016.

51 H. Yuan, J. Bi, and M. Zhou, "Profit-sensitive spatial scheduling of multi-application tasks in distributed green clouds," *IEEE Transactions on Automation Science and Engineering*, vol. 17, no. 3, pp. 1097–1106, 2020.

52 H. Yuan, J. Bi, W. Tan, M. Zhou, B. H. Li, and J. Li, "TTSA: An effective scheduling approach for delay bounded tasks in hybrid clouds," *IEEE Transactions on Cybernetics*, vol. 47, no. 11, pp. 3658–3668, 2017.

53 J. Bi, H. Yuan, W. Tan, M. Zhou, Y. Fan, J. Zhang, and J. Li, "Application-aware dynamic fine-grained resource provisioning in a virtualized cloud data center," *IEEE Transactions on Automation Science and Engineering*, vol. 14, no. 2, pp. 1172–1184, 2017.

54 H. Yuan, J. Bi, M. Zhou, and A. C. Ammari, "Time-aware multi-application task scheduling with guaranteed delay constraints in green data center," *IEEE Transactions on Automation Science and Engineering*, vol. 15, no. 3, pp. 1138–1151, 2018.

55 H. Yuan, J. Bi, and M. Zhou, "Temporal task scheduling of multiple delay-constrained applications in green hybrid cloud," *IEEE Transactions on Services Computing*, vol. 14, no. 5, pp. 1558–1570, 2021.

56 J. Robinson, S. Sinton, and Y. Rahmat-Samii, "Particle swarm, genetic algorithm, and their hybrids: Optimization of a profiled corrugated horn antenna," in *IEEE Antennas and Propagation Society International Symposium*, vol. 1. IEEE, 2002, pp. 314–317.

57 H. Mühlenbein and G. Paass, "From recombination of genes to the estimation of distributions I," in Binary Parameters. Berlin, Heidelberg: Springer-Verlag, 1996.

58 J. Ceberio, E. Irurozki, A. Mendiburu, and J. A. Lozano, "A review on estimation of distribution algorithms in permutation-based combinatorial optimization problems," *Progress in Artificial Intelligence*, vol. 1, no. 1, pp. 103–117, 2012.

59 Y.-J. Gong, J.-J. Li, Y. Zhou, Y. Li, H. S.-H. Chung, Y.-H. Shi, and J. Zhang, "Genetic learning particle swarm optimization," *IEEE Transactions on Cybernetics*, vol. 46, no. 10, pp. 2277–2290, 2016.

60 J. Li, J. Zhang, C. Jiang, and M. Zhou, "Composite particle swarm optimizer with historical memory for function optimization," *IEEE Transactions on Cybernetics*, vol. 45, no. 10, pp. 2350–2363, 2015.

61 N. Saxena, A. Tripathi, K. Mishra, and A. K. Misra, "Dynamic-PSO: An improved particle swarm optimizer," in *2015 IEEE Congress on Evolutionary Computation (CEC)*. IEEE, 2015, pp. 212–219.

62 Z. Ren, A. Zhang, C. Wen, and Z. Feng, "A scatter learning particle swarm optimization algorithm for multimodal problems," *IEEE Transactions on Cybernetics*, vol. 44, no. 7, pp. 1127–1140, 2014.

63 Q. Yang, W.-N. Chen, T. Gu, H. Zhang, J. D. Deng, Y. Li, and J. Zhang, "Segment-based predominant learning swarm optimizer for large-scale optimization," *IEEE Transactions on Cybernetics*, vol. 47, no. 9, pp. 2896–2910, 2017.

64 Q. Qin, S. Cheng, Q. Zhang, L. Li, and Y. Shi, "Particle swarm optimization with interswarm interactive learning strategy," *IEEE Transactions on Cybernetics*, vol. 46, no. 10, pp. 2238–2251, 2016.

65 S. Baskar and P. N. Suganthan, "A novel concurrent particle swarm optimization," in Congress on Evolutionary Computation, vol. 1. IEEE, 2004, pp. 792–796.

66 G. G. Yen and M. Daneshyari, "Diversity-based information exchange among multiple swarms in particle swarm optimization," *International Journal of Computational Intelligence and Applications*, vol. 7, no. 01, pp. 57–75, 2008.

67 G. G. Yen and W. F. Leong, "Dynamic multiple swarms in multiobjective particle swarm optimization," *IEEE Transactions on Systems, Man and Cybernetics, Part A: Systems and Humans*, vol. 39, no. 4, pp. 890–911, 2009.

68 H. Pan, L. Wang, and B. Liu, "Particle swarm optimization for function optimization in noisy environment," *Applied Mathematics and Computation*, vol. 181, no. 2, pp. 908–919, 2006.

69 Y. Jin and J. Branke, "Evolutionary optimization in uncertain environments-a survey," *IEEE Transactions on Evolutionary Computation*, vol. 9, no. 3, pp. 303–317, 2005.

70 A. Di Pietro, "Optimising evolutionary strategies for problems with varying noise strength," Ph.D. thesis, University of Western Australia, 2007.

71 T. Bartz-Beielstein, D. Blum, and J. Branke, "Particle swarm optimization and sequential sampling in noisy environments," in Metaheuristics. Springer, 2007, pp. 261–273.

72 J. Rada-Vilela, M. Johnston, and M. Zhang, "Population statistics for particle swarm optimization: Deception, blindness and disorientation in noisy problems," Technical Report 14-01, Victoria University of Wellington, URL: (http://ecs.victoria.ac.nz/Main/TechnicalReportSeries), Tech. Rep., 2014.

73 K. Tang, X. Li, P. N. Suganthan, Z. Yang, and T. Weise, "Benchmark functions for the CEC'2010 special session and competition on large-scale global optimization," Nature Inspired Computation and Applications Laboratory, USTC, China, 2010.

74 M. A. L. Thathachar and P. S. Sastry, "Varieties of learning automata: An overview," *IEEE Transactions on Systems, Man, and Cybernetics, Part B (Cybernetics)*, vol. 32, no. 6, pp. 711–722, 2002.

75 M. Cui, L. Li, M. Zhou, and A. Abusorrah, "Surrogate-assisted autoencoder-embedded evolutionary optimization algorithm to solve high-dimensional expensive problems," *IEEE Transactions on Evolutionary Computation*, vol. 26, no. 4, pp. 676–689, 2022.

76 Q. Kang, X. Song, M. Zhou, and L. Li, "A collaborative resource allocation strategy for decomposition-based multiobjective evolutionary algorithms," *IEEE Transactions on Systems, Man, and Cybernetics: Systems*, vol. 49, no.12, pp. 2416–2423, 2019.

77 Z. Lei, S. Gao, Z. Zhang, M. Zhou, and J. Cheng, "MO4: A many-objective evolutionary algorithm for protein structure prediction," *IEEE Transactions on Evolutionary Computation*, vol. 26, no. 3, pp. 417–430, 2022.

78 Y. Wang, S. Gao, M. Zhou, and Y. Yu, "A multi-layered gravitational search algorithm for function optimization and real-world problems," *IEEE/CAA Journal of Automatica Sinica*, vol. 8, no. 1, pp. 94–109, 2021.

79 Z. Zhao, M. Zhou, and S. Liu, "Iterated greedy algorithms for flow-shop scheduling problems: A tutorial," *IEEE Transactions on Automation Science and Engineering*, vol. 19, no. 1, pp. 251–261, 2022.

8

Applications and Future Research Directions of Learning Automata

As a powerful computational intelligence algorithm, Learning Automata (LAs) have been studied and applied widely. In this chapter, their applications and future research directions are discussed.

8.1 Summary of Existing Applications

Researchers have developed many algorithm variants of LAs to meet different application requirements. Thus LAs can be applied to real-world problems such as classification, clustering, games, knapsack problems, decision problems in networks, optimization, LA parallelization, design ranking, and scheduling.

8.1.1 Classification

Classification is the main part of supervised learning, and Support Vector Machine (SVM) is one of the most widely applied algorithms in the classification area. SVM relies upon defining a mathematical function called kernel. The kernel maps data into a high-dimensional space such that data can be separated easily. The accuracy of SVM depends on the correct choice of kernel. It is difficult to choose a correct kernel because the number of kernels is infinite. Goodwin and Yazidi propose an LA-based PolyLA for classification [1]. PolyLA deals with classification problems in two-dimensional Euclidean matrices by building "separators" with polygons, and it does not involve the use of kernels. Each node in the grid maps to LA whose actions select the edges of a polygon separator. The polygon represents a path that connects head to tail. In each iteration, the LA selects a polygon randomly based on the distribution of a set of possible paths. The observed classification accuracy is used to strengthen the polygon to increase

Learning Automata and Their Applications to Intelligent Systems, First Edition.
JunQi Zhang and MengChu Zhou.
© 2024 The Institute of Electrical and Electronics Engineers, Inc. Published 2024 by John Wiley & Sons, Inc.

the probability of selecting it again. Paths that produce low performance receive weaker enhancement signals, and they are thus selected less frequently. The polygon where the best path lies is strengthened, and the actions of LA converge to a polygon forming a good separation. Thus, a self-enclosing polygon splitter is formed and generated. The generated polygon separators encapsulate all items of each class. Compared with SVM, the performance of PolyLA is more stable because it does not depend on the selection and use of kernels.

In order to improve the accuracy of classification and shorten the time of classification, Afshar *et al.* propose a new LA-based classifier named MLAC for multi-class classification [2]. In the proposed algorithm, a solution space is divided into several hypercubes, and each hypercube is mapped to an action of LA. Each hypercube has a uniformly initialized probability vector. The probability of each hypercube is the probability that this hypercube is selected. Then, these hypercubes and their corresponding probabilities are modified gradually in an iterative process until the volumes of all hypercubes are smaller than the set minimum value. Hypercubes with the highest selection probability become candidate classification results. Experimental results show that MLAC has higher accuracy and shorter running time than traditional classifiers such as K-Nearest Neighbor (KNN), Multilayer Perceptron (MLP), Genetic Algorithm (GA)-classifier, and Particle Swarm (PS)-classifier [3, 4]. More comparison work with some recent methods, e.g., [5–12] should be performed.

Many Learning Classifier Systems (LCSs) are designed to learn state-action value functions through a set of general and precise rules for solving reinforcement learning problems. Most of these systems employ greedy action-selection strategies to learn deterministic strategies. In a fully observable and deterministic environment, there exists at least one optimal deterministic policy that can be obtained by using a greedy policy. Different from fully observable environments, partially observable environments lack Markov properties. Due to this limitation, the learning agent may be trapped into local loops or local optima and fail to achieve its learning goals when deterministic policies are employed. Compared to learning deterministic policies, agents can easily jump out of local loops by learning stochastic policies. Based on the above idea, Chen *et al.* propose a new mechanism to learn random action-selection policies [13]. In this mechanism, each classifier of LCS contains a new policy parameter that controls the probability of executing each action directly. The prediction parameters can be removed safely without any performance degradation when testing the performance of the classifier. LCS can learn any stochastic policies by using policy parameters and treat deterministic policies as a special stochastic policy. The performance of LCS is improved effectively because of adaptive learning of action selection policies, especially in stochastic and partially observable learning environments.

8.1.2 Clustering

Similar to classification, clustering of data has been an important research content for a long time in machine learning research field. Hasanzadeh-Mofrad and Rezvanian propose an LA clustering (LAC) [14] algorithm. In LAC, each data point belongs to LA that determines the cluster membership for that data point. The specific algorithm flow is as follows: First, assign an LA to each data point. Secondly, each LA determines the cluster membership of data points. The data samples that the LA tries to correspond to are placed in the most coherent clusters, which can be intuitively mapped to the LA's action set. Each action represents selecting a specific cluster for the corresponding data sample. Then, the new centroids (means) are calculated. Finally, the reinforcement signal for each LA is calculated by comparing the new centroid with the previous one, and the cluster assignments are adjusted according to the reinforcement signal. The experimental results show that LAC has higher clustering accuracy and comparable Silhouette coefficient compared with K-means, K-means++, K-median, and other clustering algorithms [15].

Determining the number of clusters has been one of the main challenges in the clustering field. Anari *et al.* propose a new dynamic clustering method based on Continuous Action-set LA (CALA) named ACCALA [16]. CALA is an optimization tool that interacts with a stochastic environment and learns the optimal operations from environment. In ACCALA, a set of CALAs find the accurate number of clusters and their centers. ACCALA regards the dynamic clustering problem as a noisy optimization problem and treats the clusters and their centers as the parameter space of the problem. Based on it, the model and a set of CALAs are built. According to the dataset, the set of actions for each LA is defined on a continuous scale. After building a set of CALAs, each automaton in the set selects an action from its successive set of actions randomly and independently. The chosen action is applied to the random environment that feedbacks reinforcement signals to the automata. Each automaton updates its parameters according to the reinforcement signal. ACCALA repeats this learning process until the minimum value of the optimization function is reached. After this, the number of clusters and their centers are determined. Experimental results show that ACCALA is able to find the accurate number of clusters and is more efficient than other clustering methods. The appropriate number of segments can be found automatically by applying the method to the image segmentation problem. More comparison work between it and recently proposed clustering methods, e.g., [17–19], is needed.

8.1.3 Games

Game theory provides an ideal environment for studying multiplayer optimal decision-making and control problems. LA has gained some applications in the field of game theory. Many algorithms utilize LA to converge to the Nash equilibrium of the game with limited information. Most LA algorithms for games are absorbed in the probabilistic simplex space, and they converge to the exclusive choice of a single action. So these LA cannot converge to other mixed Nash equilibria when the pure policy has no saddle points. Yazidi *et al.* propose a solution for LA: If a pure policy has a saddle point, the scheme converges to a near-optimal solution close to the pure policy in the probabilistic simplex; even using pure policies, convergence to an optimal mixed Nash equilibrium may not have a saddle-point LA. Yazidi *et al.* have proved the theoretical results of the algorithm's convergence and stability [20].

Markov decision models allow a single agent to learn a policy that maximizes potentially delayed reward signals in a stochastic stationary environment. Convergence to the optimal policy is guaranteed as long as the agent can fully explore the environment in which it operates and owns Markov properties. But the Markov decision model does not allow multiple agents to operate in the same environment. Using a Markov game framework, Vrancx *et al.* give a direct extension of the Markov decision model to multiple agents [21]. In a Markov game, actions are the result of joint action choices of all agents. The rewards and state transitions depend on these joint actions. For each agent, there is a simple LA in each state. This setup can be analyzed from three different perspectives: A single superagent view, where a single agent is represented by the entire automaton set; a multi-agent view, where each agent is represented by an automaton for each associated state; LA views, where each automaton represents an agent. Vrancx *et al.* prove that a set of independent LA can reach a pure equilibrium point in Markov games.

8.1.4 Knapsack Problems

Resource allocation and scheduling are often involved in real world, such as resource allocation in web monitoring and real-time allocation of limited sampling resources. Granmo *et al.* model the above problem as a knapsack problem and propose a single general solution [22]. Based on this model, a set of LAs is used to perform random walks on the discrete fraction space to solve the knapsack problems. Comprehensive experimental results show that the discretization resolution determines the accuracy of the scheme. For a given accuracy, LA can improve the current solution continuously until a near-optimal solution is found. The above-mentioned type of resource allocation problems can be solved effectively by this method.

Resource allocation is often based on incomplete and noisy information in real world. Incomplete and noisy information in this resource allocation problem causes traditional optimization techniques ineffective. This type of problems can be formulated as a Stochastic Nonlinear Fractional Equality Knapsack (NEFK) problem. Granmo and Oommen propose a component named Twofold Resource Allocation Automaton (TRAA) [23]. Based on TRAA, a new online LA system, namely the Hierarchy of Twofold Resource Allocation Automaton (H-TRAA), is proposed to solve this type of knapsack problems. In H-TRAA, each TRAA works with two resources and moves randomly along a discrete probability space. Comprehensive experimental results show that H-TRAA outperforms previous state-of-the-art schemes like learning automata knapsack game [24].

8.1.5 Decision Problems in Networks

With the rapid growth in the number of mobile user terminals and the demand for mobile bandwidth services, wireless data traffic is expected to continue to increase. Among all wireless data traffic, video-on-demand data traffic occupies a large proportion. In video-on-demand, redundant transmission of download requests is common. To reduce the redundancy of data transfer, an efficient solution is to store popular data locally into the helpers, called caching. A significant advantage of caching is the reduction of download delays due to short-range communication between the helper and the user terminal. How to put data into helpers to minimize download delay is an important problem, and the problem is proved to be NP-hard. Marini *et al.* propose a distributed and decentralized algorithm to optimize the placement of cached data in heterogeneous cellular networks [25]. The algorithm is based on a set of LAs that minimize user-perceived latency by coordinating an independent team of LA to collectively focus on data placement. Aiming at the rapid convergence of a single LA, the algorithm introduces the concept of conditional inaction and proposes a discrete generalized pursuit algorithm, and designs a frequent suitable reward function for each LA. Statistical studies have shown that the distributed solution is close to the performance of the centralized greedy strategy, while exhibiting low computational complexity.

With the increasing popularity of new wireless technologies and increasing demand for spectrum, the shortcomings of traditional command-and-control based static spectrum allocation methods are exposed. Cognitive Radio Networks (CRNs) have emerged as a promising solution to the problems of underutilized radio spectrum shortage. The dynamic spectrum access paradigm allows for the coexistence of a secondary network consisting of Secondary Users (SUs) and a primary network consisting of licensed Primary Users (PUs), but it should ensure that the SUs do not cause any interference to the primary network. This means that SUs should free up channels when PUs arise, and SUs should only be accessible when spectrum opportunities arise. Because of the dynamic nature of CRNs and the fact that an SU may not share a common channel with all of its neighbors, the task of multicast routing is challenging. To solve this problem, Ali *et al.* propose an on-demand multicast protocol named Multipath on-Demand Multicast

Routing (MP-ODMR) [26]. The protocol combines the ability to use multiple paths and the advantage of the availability of multiple radios to improve routing performance. In CRN, efficient routing requires incorporating learning and spectrum prediction into the routing framework. Therefore the multicast routing problem is formulated in this protocol as an LA problem. LA is used to identify the best path and best channel. LA punishes or rewards routes based on successful receipt of ACK-response. Assuming that PUs have no mobility, a particular PU affects one link and does not affect other links. So as long as this particular PU is active, the corresponding link is not to be selected and the probability of selecting other links increases. The links where the PU has no influence are the optimal paths for transmission, and these optimal paths are learned by the LA. After a period of time, the probability of a large number of successful transmission routes and channels being selected becomes higher, and the best routes and channels for data transmission can thus be screened out. Simulation results show that the multi-path multicast approach is feasible, and the proposed protocol shows better performance than the state-of-the-art CRN multicast protocol.

As next-generation wireless access infrastructure, broadband wireless access systems are gaining popularity. The Worldwide Interoperability for Microwave Access (WiMAX) has been a family of standards for wireless metropolitan area networks and is presently used in some countries or rural areas despite it has been mostly replaced by more advanced technologies such as Long-Term Evolution (LTE) and 5G. The WiMAX access network can determine the ratio between the downstream and upstream directions flexibly by introducing a flexible, efficient, and robust radio interface in WiMAX. Sarigiannidis *et al.* propose an adaptive model, which attempts to fully adjust the subframe width ratio from downlink to uplink according to the current traffic [27]. This model can solve this problem effectively. In the context of mobile WiMAX radio access networks, the base station is augmented by error-aware LA that is able to identify the magnitude of incoming and outgoing traffic and then define the ratio appropriately on a frame-by-frame basis. The proposed model takes into account the downstream and upstream feedback obtained from the mapping process and applies an empirical learning framework to adopt the traffic demand fully. As a learning tool, LA is used to process feedback obtained and actions recorded in the past. The model constitutes a dynamic and adaptive downlink–uplink ratio adjustment scheme based on short-term network dynamics. The designed model is evaluated extensively in realistic and dynamic scenarios. The results show that the performance is significantly improved over the scheme with predefined and fixed ratios.

8.1.6 Optimization

Many stochastic search techniques are very sensitive to parameter settings. Similar to most stochastic search techniques, modifying a single parameter in a Particle

Swarm Optimizer (PSO) can have large impact. Therefore, an efficient parameter setting is an important part of the stochastic search technique. Hashemi and Meybodi propose an adaptive PSO parameter selection method based on LA for adaptive adjustment of PSO parameters, inertia weights, cognitive and social parameters [28]. In this method, three LAs are employed, one for each parameter setting. There are two kinds of adjustment strategies for parameter values: In the first strategy, parameter values are chosen from a limited set; in the second one, the LA decides conservatively to change the parameter value by a fixed amount or make no change. Experimental results show that the proposed algorithm can reach the optimization goal faster than most other PSO algorithm variants for all benchmark functions.

PSO and most of its variants suffer from getting stuck in local optima. Because of this, Hasanzadeh *et al.* propose a Dynamic Global and Local Combined Particle Swarm Optimization based on a three-action LA (DPSOLA) [29]. DPSOLA divides particle swarms into sub-populations in which particles can share their information. The embedded LA collects information from individuals, local optimal particles, and global optimal particles. Such information is used to navigate the particles in the target space. In each iteration, LA controls the trend of particle search by selecting an optimal action from a limited set of actions. In order to reduce the probability that LA loses the useful information of the particles, each particle is assigned an LA to save the useful information and improve the diversity of the particle swarm. The algorithm has been tested on benchmark functions in different dimensions. The results show that both fitness and convergence speed are superior to the traditional PSO and its variants.

Multi-objective optimization problems require weighing conflicting objectives to obtain a set of solutions. Heuristic search can provide effective cross-domain search methods applicable to different problems. There are some studies that combine the advantages of multiple multi-objective evolutionary algorithms to provide better overall performance for multi-objective optimization. Li *et al.* propose a new LA-based hyperheuristic selection framework, which implements LA-based Hyper-heuristic (HHLA) and LA-based Hyper-heuristic with Ranking Scheme Initialization (HH-RILA) for multi-objective optimization [30]. When solving a given problem, LA can guide the selection of an appropriate multi-objective evolutionary algorithm at each decision point. The two proposed alternative hyperheuristics differ in the initial setup process. HH-LA takes all three low-level multi-objective evolutionary algorithms and gives each algorithm an equal chance to start randomly. HH-RILA adopts a ranking scheme that eliminates the relatively poor performing multi-objective evolutionary algorithms and uses the remaining multi-objective evolutionary algorithms in the improvement process. The performance of the proposed hyperheuristic selection framework is tested on benchmark functions and vehicle crashworthiness problems. Empirical results show that the proposed hyperheuristic algorithm is effective and versatile.

In order to improve the search efficiency of multi-objective evolutionary algorithms and maintain the diversity of solutions, Dai *et al.* use LA for Quantization Orthogonal Crossover (QOX) and propose a new decomposition-based fitness function [31]. The algorithm employs LA, QOX, and a new fitness function. LA is used to perform mutation operators and to group the decision variables of QOX. Improved QOX is more likely to produce good offspring by using historical information to group variables to improve search efficiency. Furthermore, a fitness function based on the decomposition of the target space is proposed to preserve the diversity and uniformity of the solutions and guide the obtained solutions to approach the true Pareto front. The experimental results show that the algorithm can perform a good search in a continuous region. It can effectively achieve a more accurate and wider Pareto front than its compared multiobjective optimization algorithms.

For continuous function optimization, Zeng and Liu propose a new optimization method using LA [32]. The basic idea of this method is as follows: First, the value range of an optimization function is divided into several intervals evenly, each interval corresponds to an action of LA. The estimates of the probability of action and the probability of reward for each cycle are updated by computing the value of the function for a selected sample corresponding to the current action randomly. The estimate of the reward probability is compared with a predefined threshold, and then the corresponding interval based on a sample of the calculated function values is evaluated. If the mean and variance of these function values are small enough, the interval is deleted because the interval is considered stable and useless. Otherwise, the interval is unstable, and the change of the function in this interval is estimated from the samples in it. Next, this interval is divided into several subintervals and other intervals are removed. This process is repeated until a predefined accuracy condition is met, and the original interval is removed or converged to several values that contain a global optimum. This approach reduces search in useless areas. The method takes into account optimization accuracy and experimental cost, and it allows one to obtain values that are quite close to the global optimum with a small number of samples.

8.1.7 LA Parallelization and Design Ranking

In the theoretical field, LA networks based on the synthesis of several LAs are devoted to solving problems that are intractable by a single LA. Their purpose is to improve the convergence speed by taking advantage of the parallel nature of the environment. The learning process of a single LA is sequential in nature, choosing only one action at a time and eliciting one feedback from the environment. In some cases the environment can respond to multiple actions simultaneously. Multiple actions can then be sent together as one input, and all feedback signals can be used together to update the action probabilities. LA networks have been proved to combine several LA into a system efficiently, resulting in faster convergence than a single LA. The rationale behind this is that if the learning processes of the agent in the system are not correlated strongly, then the agent can learn different aspects of a problem. What agents have in common can reinforce learning outcomes. Inspired by this, Ge *et al.* propose a new parallel LA network framework, expecting this parallel system to have the characteristics of learning agents [33]. A parallel operation is divided into two steps: decentralized learning and centralized fusion. The number of interactions required between the merged pursuit LA and the environment is reduced by using decentralized learning and centralized fusion. Various aspects of current issues are learnt and what has been learned through intensive fusion is summarized through diversity by using decentralized learning. A decentralized learning process uses a deterministic estimator based on LA, and uses trial-and-error information to refine the probability vector. During the centralized fusion process, the updated probability vectors are combined into a common probability. In the proposed framework, the averaging strategy is used as the default operation in the centralized fusion stage. For each component in the common vector, the averaging strategy is to average the temporal probability values of all LAs at that location. Simulation results verify the effectiveness of the parallel framework and demonstrate its superiority to single-LA one. The framework is further applied to a random point localization problem and obtains high-quality performance.

Due to the large amount of heterogeneous information on the network, the information retrieval process needs to deal with a large number of uncertainties and doubts. In this case, it is crucial to design an efficient retrieval function and sorting algorithm to provide the most relevant results. Akbari Torkestani propose a sorting function discovery algorithm based on LA [34]. In this algorithm, LA is responsible for calculating the average importance of each information source in the final ranking. All sources of information have equal importance in the initial stage. All actions are chosen with the same probability. If a user tries to view a document, the document is more likely to be relevant to the searched topic. Actions corresponding to the source of the reviewed document are rewarded. For each query, the proposed algorithm increases the portion where the action of the information source is rewarded and reduces the action of other sources. In the final ranking, the weight of reward evidence sources is increased. As the algorithm progresses, the probability of each information source appearing in the final ranking is proportional to its relevance to the user's query. The algorithm is tested on different known data sets for several user query types. The obtained results are compared with some existing methods. The results show that the algorithm is superior to its peers in terms of accuracy and normalized discount cumulative gain.

To solve the problem of information overload on the Internet, recommender systems [35–41] enable users to obtain resources that best meet their needs and interests by guiding users to extract knowledge from previous user interactions. Forsati and Meybodi use distributed LA to learn the behavior of previous users and recommend pages to the current user based on the learned patterns [42]. The algorithm leverages web usage data and the underlying site structure to recommend pages to the current user. In this algorithm, LA learns a transition probability matrix from the behavior of visiting users available in the site's log files. In addition, the personalization vector is calculated based on the visit rate of the preference pages visited by the user. Besides page recommendation, the algorithm can also be used to modify links among pages.

8.1.8 Scheduling

The development of cloud computing meets the different customer needs. However, cloud computing systems consume a lot of power while providing services, resulting in higher operating cost. Efficient task scheduling is an effective way to ensure that cloud users are satisfied with deadline-sensitive services while minimizing energy consumption [43–47]. Sahoo *et al.* formulate the task scheduling problem as a dual-objective minimization problem involving energy minimization and time maximization [48]. They propose a new LA-based scheduling framework to solve the problem. LA in the proposed framework is an adaptive decision-making unit that helps the scheduler choose the optimal action. Reinforcement learning schemes for LA are used to obtain optimal actions from a set of possible actions. The framework introduces a scheduling algorithm based on LA. The algorithm utilizes the heterogeneity of tasks and virtual machines to ensure the timing requirements of tasks. The experimental results show that the scheduling framework has advantages in energy consumption, maximum time, and success rate. The framework has broad applicability for deadline-sensitive task scheduling in heterogeneous cloud environments [49–52].

Grid computing is a new field of distributed computing focused on resource sharing. Existing grid resource discovery solutions do not adapt well to dynamic and heterogeneous grids. Hasanzadeh and Meybodi propose a grid resource discovery method for distributed LAs [53]. The method utilizes a network of distributed LAs for forwarding domain-specific queries and constructs a multi-layer architecture of the relationship between grid entities and distributed LAs. The bottom resource layer starts from the various grid resources available, and then the virtual organization layer enforces local policies on all member resources. The grid information system layer stores and organizes the information of virtual organization member resources, and the distributed LA layer performs grid resource discovery. Finally, the top user layer provides an engine to communicate with the grid virtual organization and initiate resource requests. Analysis and experiments show that the method can cope with grid changes adaptively. Furthermore, the method has the best performance in terms of hop count, query hit count, and resource utilization on small, medium, and large grid scales among all compared methods. Especially in small-scale grids, their algorithm is more accurate than large-scale grids, although the action set of distributed LA is small.

8.2 Future Research Directions

The following parts analyze and discuss four aspects in research trends of LA: parameter setting optimization, spatial environment expansion, combination with hyperheuristic optimization methods, and combination with multi-agent reinforcement learning.

LA learns the optimal action based on the interaction between intelligent agents and random environments. It is an important tool for reinforcement learning. For unknown environments, most LAs have multiple parameters that need to be tuned during pre-training for interacting with the environment. Only when the parameters are properly tuned can the LA act most correctly during the training process to obtain the best behavior. Most parameters need to be tuned for each individual random environment to stabilize the performance of LA. Furthermore, the interaction time required for parameter tuning can be enormous. These parameters are only adjusted by additional interaction with the environment. This can be unbearable in practical applications, especially if the interaction is expensive [54]. Therefore it is crucial that LA works well in all stochastic environments without environment-dependent parameter tuning.

The field of LA has been studied and analyzed extensively for many years. LA is a well-established computational model which is suitable for the optimal machine design in stationary and non-stationary environments. Most of the research has focused on LA to work in environments with a limited number of actions, and all these algorithms assume that the reward function has a well-defined and very strict functional form. At present, most LAs can only deal with environments with a limited number of actions, and only a few continuous action LA can deal with spatial environments with continuous actions. These LAs require very strict assumptions about the reward function, which are difficult to satisfy in some practical problems. Lu define an environment with a continuous action space: First, the environment for LA has infinite actions. Second, the unimodal functional form of the infinite reward probability of the environment does not obey a perfect form. Third, potential functional forms include, but not limited to, multimodal functions. Fourth, all solutions further move the search field in a stepwise fashion, thus encompassing the field of discrete LA. Lu *et al.* propose a continuous action LA to analyze and discuss this problem fully, and show the necessity of fully researching this problem [55].

As mentioned earlier, metaheuristics tailored for specific domains have been applied to many computationally difficult optimization problems successfully [30]. In the most of existing studies, a single metaheuristic is designed and applied to a specific problem. However, their application to a new problem domain often requires additional expert intervention. Hyper-heuristic Selection is one of the main research directions for cross-domain search methods, which require only a small amount of expert intervention and can be applied to problem instances from different domains. Improving the generality level of heuristic optimization methods is one of the main motivations for hyperheuristic research, which provides a general heuristic optimization framework for taking advantage of multiple heuristics. The idea is to provide an effective and reusable cross-domain search method through the automation of heuristic search. Hence, the heuristic method can be applied to problems with different characteristics in different fields and does not require too many experts to participate. Learning is key to developing adaptive hyperheuristics for effective choice. LA can guide users to choose appropriate heuristics at each decision point when solving a given problem. The aforementioned LA-based selection hyperheuristics demonstrate the research potential in this area.

Multi-agent reinforcement learning has been used widely in many applications due to its ease of implementation and distributed tasks [56–61]. In a single-agent setting, the agent receives a numerical reward from the environment after performing an action and uses the reward to improve its behavior for the maximum expected reward. However in a multi-agent setting, ensemble learning becomes infeasible due to the exponential growth of the joint action space and limited sensing range. Some multi-agent reinforcement learning algorithms with independent learners do not require any agent to observe the behavior of other agents. This type of multi-agent reinforcement learning algorithm can drastically reduce communications among agents, alleviating the problem of growing joint action spaces. As an independent learner in multi-agent reinforcement learning, LA has the following advantages: First, LA has all the advantages of an independent learner; second, LA requires a simple reinforcement signal from the environment; third, LA requires simple mathematical operations in each time step, and it is thus suitable for real-time applications; fourth, LA is simple in its structure and easy to implement. Based on the above advantages, LA is appropriate to obtain optimal joint action or some type of balance. Many LA-based algorithms use more than one LA [62]. Therefore, the combination of LAs and multi-agent reinforcement learning [63–65] has important research and application value.

8.3 Exercises

1 Combined with the above content, summarize the role of LA in the above application fields.

2 Point out the shortcomings of LA in the above application scenarios, and try to propose solutions to improve these shortcomings.

3 What other application scenarios exist for LA besides the above application scenarios?

4 Combined with the content of the text and the algorithm principle of LA, summarize the research trends in the field of LA.

5 In addition to the content mentioned in the text, what are the feasible research directions in the field of LA research?

6 Give an example of an real application and use LA to improve its performance.

7 What are the kind of applications where LA can help?

References

1 M. Goodwin and A. Yazidi, "Distributed learning automata-based scheme for classification using novel pursuit scheme," *Applied Intelligence*, vol. 50, no. 7, pp. 2222–2238, 2020.

2 S. Afshar, M. Mosleh, and M. Kheyrandish, "Presenting a new multiclass classifier based on learning automata," *Neurocomputing*, vol. 104, pp. 97–104, 2013.

3 W. Dong and M. Zhou, "Gaussian classifier-based evolutionary strategy for multimodal optimization," *IEEE Transactions on Neural Networks and Learning Systems*, vol. 25, no. 6, pp. 1200–1216, 2014.

4 S. Gao, M. Zhou, Y. Wang, J. Cheng, H. Yachi, and J. Wang, "Dendritic neuron model with effective learning algorithms for classification, approximation, and prediction," *IEEE Transactions on Neural Networks and Learning Systems*, vol. 30, no. 2, pp. 601–614, 2019.

5 W. Zhang, H. Zhang, J. Liu, K. Li, D. Yang, and H. Tian, "Weather prediction with multiclass support vector machines in the fault detection of photovoltaic system," *IEEE/CAA Journal of Automatica Sinica*, vol. 4, no. 3, pp. 520–525, 2017.

6 Q. Kang, L. Shi, M. Zhou, X. Wang, Q. Wu, and Z. Wei, "A distance-based weighted undersampling scheme for support vector machines and its application to imbalanced classification," *IEEE Transactions on Neural Networks and Learning Systems*, vol. 29, no. 9, pp. 4152–4165, 2018.

7 Q. Kang, X. Chen, S. Li, and M. Zhou, "A noise-filtered under-sampling scheme for imbalanced classification," *IEEE Transactions on Cybernetics*, vol. 47, no. 12, pp. 4263–4274, 2017.

8 P. Zhang, S. Shu, and M. Zhou, "An online fault detection method based on SVM-grid for cloud computing systems," *IEEE/CAA Journal of Automatica Sinica*, vol. 5, no. 2, pp. 445–456, 2018.

9 H. Liu, M. Zhou, and Q. Liu, "An embedded feature selection method for imbalanced data classification," *IEEE/CAA Journal of Automatica Sinica*, vol. 6, no. 3, pp. 703–715, 2019.

10 H. Zhu, G. Liu, M. Zhou, Y. Xie, A. Abusorrah, and Q. Kang, "Optimizing weighted extreme learning machines for imbalanced classification and application to credit card fraud detection," *Neurocomputing*, vol. 407, pp. 50–62, 2020.

11 X. Luo, X. Wen, M. Zhou, A. Abusorrah, and L. Huang, "Decision-tree-initialized dendritic neuron model for fast and accurate data classification," *IEEE Transactions on Neural Networks and Learning Systems*, vol. 33, no. 9, pp. 4173–4183, 2022.

12 Z. Tan, J. Chen, Q. Kang, M. Zhou, A. Abusorrah, and K. Sedraoui, "Dynamic embedding projection-gated convolutional neural networks for text classification," *IEEE Transactions on Neural Networks and Learning Systems*, vol. 33, no. 3, pp. 973–982, 2022.

13 G. Chen, C. I. Douch, and M. Zhang, "Using learning classifier systems to learn stochastic decision policies," *IEEE Transactions on Evolutionary Computation*, vol. 19, no. 6, pp. 885–902, 2015.

14 M. Hasanzadeh-Mofrad and A. Rezvanian, "Learning automata clustering," *Journal of Computational Science*, vol. 24, pp. 379–388, 2018.

15 X. S. Lu, M. Zhou, L. Qi, and H. Liu, "Clustering-algorithm-based rare-event evolution analysis via social media data," *IEEE Transactions on Computational Social Systems*, vol. 6, no. 2, pp. 301–310, 2019.

16 B. Anari, J. A. Torkestani, and A. M. Rahmani, "Automatic data clustering using continuous action-set learning automata and its application in segmentation of images," *Applied Soft Computing*, vol. 51, pp. 253–265, 2017.

17 M. Ghahramani, A. O'Hagan, M. Zhou, and J. Sweeney, "Intelligent geodemographic clustering based on neural network and particle swarm optimization," *IEEE Transactions on Systems, Man, and Cybernetics: Systems*, vol. 52, no. 6, pp. 3746–3756, 2022.

18 X. Xu, J. Li, M. Zhou, J. Xu, and J. Cao, "Accelerated two-stage particle swarm optimization for clustering not-well-separated data," *IEEE Transactions on Systems, Man, and Cybernetics: Systems*, vol. 50, no. 11, pp. 4212–4223, 2020.

19 T. Zhang, M. Zhou, X. Guo, L. Qi, and A. Abusorrah, "A density-center-based automatic clustering algorithm for iot data analysis," *IEEE Internet of Things Journal*, vol. 9, no. 24, pp. 24682–24694, 2022.

20 A. Yazidi, D. Silvestre, and B. J. Oommen, "Solving two-person zero-sum stochastic games with incomplete information using learning automata with artificial barriers," *IEEE Transactions on Neural Networks and Learning Systems*, vol. 34, no. 2, pp. 650–661, 2021.

21 P. Vrancx, K. Verbeeck, and A. Nowé, "Decentralized learning in Markov games," *IEEE Transactions on Systems, Man, and Cybernetics, Part B (Cybernetics)*, vol. 38, no. 4, pp. 976–981, 2008.

22 O.-C. Granmo, B. J. Oommen, S. A. Myrer, and M. G. Olsen, "Learning automata-based solutions to the nonlinear fractional knapsack problem with applications to optimal resource allocation," *IEEE Transactions on Systems, Man, and Cybernetics, Part B (Cybernetics)*, vol. 37, no. 1, pp. 166–175, 2007.

23 O.-C. Granmo and B. J. Oommen, "Solving stochastic nonlinear resource allocation problems using a hierarchy of twofold resource allocation automata," *IEEE Transactions on Computers*, vol. 59, no. 4, pp. 545–560, 2010.

24 B. J. Oommen, "Stochastic searching on the line and its applications to parameter learning in nonlinear optimization," *IEEE Transactions on Systems, Man, and Cybernetics, Part B (Cybernetics)*, vol. 27, no. 4, pp. 733–739, 1997.

25 L. Marini, J. Li, and Y. Li, "Distributed caching based on decentralized learning automata," in *2015 IEEE International Conference on Communications (ICC)*. IEEE, 2015, pp. 3807–3812.

26 A. Ali, J. Qadir, and A. Baig, "Learning automata based multipath multicasting in cognitive radio networks," *Journal of Communications and Networks*, vol. 17, no. 4, pp. 406–418, 2015.

27 A. G. Sarigiannidis, P. Nicopolitidis, G. I. Papadimitriou, P. G. Sarigiannidis, M. D. Louta, and A. S. Pomportsis, "On the use of learning automata in tuning the channel split ratio of WiMAX networks," *IEEE Systems Journal*, vol. 9, no. 3, pp. 651–663, 2013.

28 A. B. Hashemi and M. Meybodi, "Adaptive parameter selection scheme for PSO: A learning automata approach," in *2009 14th International CSI Computer Conference*. IEEE, 2009, pp. 403–411.

29 M. Hasanzadeh, M. R. Meybodi, and S. S. Ghidary, "Improving learning automata based particle swarm: An optimization algorithm," in *2011 IEEE 12th International Symposium on Computational Intelligence and Informatics (CINTI)*. IEEE, 2011, pp. 291–296.

30 W. Li, E. Özcan, and R. John, "A learning automata-based multiobjective hyper-heuristic," *IEEE Transactions on Evolutionary Computation*, vol. 23, no. 1, pp. 59–73, 2017.

31 C. Dai, Y. Wang, M. Ye, X. Xue, and H. Liu, "An orthogonal evolutionary algorithm with learning automata for multiobjective optimization," *IEEE Transactions on Cybernetics*, vol. 46, no. 12, pp. 3306–3319, 2015.

32 X. Zeng and Z. Liu, "A learning automata based algorithm for optimization of continuous complex functions," *Information Sciences*, vol. 174, no. 3–4, pp. 165–175, 2005.

33 H. Ge, J. Li, S. Li, W. Jiang, and Y. Wang, "A novel parallel framework for pursuit learning schemes," *Neurocomputing*, vol. 228, pp. 198–204, 2017.

34 J. Akbari Torkestani, "An adaptive learning automata-based ranking function discovery algorithm," *Journal of Intelligent Information Systems*, vol. 39, no. 2, pp. 441–459, 2012.

35 X. Luo, M. Zhou, Y. Xia, and Q. Zhu, "An efficient non-negative matrix-factorization-based approach to collaborative filtering for recommender systems," *IEEE Transactions on Industrial Informatics*, vol. 10, no. 2, pp. 1273–1284, 2014.

36 X. Luo, M. Zhou, S. Li, Z. You, Y. Xia, and Q. Zhu, "A nonnegative latent factor model for large-scale sparse matrices in recommender systems via alternating direction method," *IEEE Transactions on Neural Networks and Learning Systems*, vol. 27, no. 3, pp. 579–592, 2016.

37 X. Luo, M. Zhou, S. Li, D. Wu, Z. Liu, and M. Shang, "Algorithms of unconstrained non-negative latent factor analysis for recommender systems," *IEEE Transactions on Big Data*, vol. 7, no. 1, pp. 227–240, 2021.

38 X. Luo, W. Qin, A. Dong, K. Sedraoui, and M. Zhou, "Efficient and high-quality recommendations via momentum-incorporated parallel stochastic gradient descent-based learning," *IEEE/CAA Journal of Automatica Sinica*, vol. 8, no. 2, pp. 402–411, 2021.

39 G. Wei, Q. Wu, and M. Zhou, "A hybrid probabilistic multiobjective evolutionary algorithm for commercial recommendation systems," *IEEE Transactions on Computational Social Systems*, vol. 8, no. 3, pp. 589–598, 2021.

40 D. Wu, X. Luo, M. Shang, Y. He, G. Wang, and M. Zhou, "A deep latent factor model for high-dimensional and sparse matrices in recommender systems," *IEEE Transactions on Systems, Man, and Cybernetics: Systems*, vol. 51, no. 7, pp. 4285–4296, 2021.

41 W. Yue, Z. Wang, J. Zhang, and X. Liu, "An overview of recommendation techniques and their applications in healthcare," *IEEE/CAA Journal of Automatica Sinica*, vol. 8, no. 4, pp. 701–717, 2021.

42 R. Forsati and M. R. Meybodi, "Effective page recommendation algorithms based on distributed learning automata and weighted association rules," *Expert Systems with Applications*, vol. 37, no. 2, pp. 1316–1330, 2010.

43 B. Hu, Z. Cao, and M. Zhou, "Scheduling real-time parallel applications in cloud to minimize energy consumption," *IEEE Transactions on Cloud Computing*, vol. 10, no. 1, pp. 662–674, 2022.

44 H. Yuan, J. Bi, J. Zhang, and M. Zhou, "Energy consumption and performance optimized task scheduling in distributed data centers," *IEEE Transactions on Systems, Man, and Cybernetics: Systems*, vol. 52, no. 9, pp. 5506–5517, 2022.

45 H. Yuan, M. Zhou, Q. Liu, and A. Abusorrah, "Fine-grained resource provisioning and task scheduling for heterogeneous applications in distributed green clouds," *IEEE/CAA Journal of Automatica Sinica*, vol. 7, no. 5, pp. 1380–1393, 2020.

46 H. Yuan, J. Bi, M. Zhou, and A. C. Ammari, "Time-aware multi-application task scheduling with guaranteed delay constraints in green data center," *IEEE Transactions on Automation Science and Engineering*, vol. 15, no. 3, pp. 1138–1151, 2018.

47 H. Yuan, H. Liu, J. Bi, and M. Zhou, "Revenue and energy cost-optimized biobjective task scheduling for green cloud data centers," *IEEE Transactions on Automation Science and Engineering*, vol. 18, no. 2, pp. 817–830, 2021.

48 S. Sahoo, B. Sahoo, and A. K. Turuk, "A learning automata-based scheduling for deadline sensitive task in the cloud," *IEEE Transactions on Services Computing*, vol. 14, no. 6, pp. 1662–1674, 2019.

49 H. Yuan and M. Zhou, "Profit-maximized collaborative computation offloading and resource allocation in distributed cloud and edge computing systems," *IEEE Transactions on Automation Science and Engineering*, vol. 18, no. 3, pp. 1277–1287, 2021.

50 H. Yuan, J. Bi, and M. Zhou, "Multiqueue scheduling of heterogeneous tasks with bounded response time in hybrid green iaas clouds," *IEEE Transactions on Industrial Informatics*, vol. 15, no. 10, pp. 5404–5412, 2019.

51 H. Yuan, J. Bi, and M. Zhou, "Spatiotemporal task scheduling for heterogeneous delay-tolerant applications in distributed green data centers," *IEEE Transactions on Automation Science and Engineering*, vol. 16, no. 4, pp. 1686–1697, 2019.

52 Q.-H. Zhu, H. Tang, J.-J. Huang, and Y. Hou, "Task scheduling for multi-cloud computing subject to security and reliability constraints," *IEEE/CAA Journal of Automatica Sinica*, vol. 8, no. 4, pp. 848–865, 2021.

53 M. Hasanzadeh and M. R. Meybodi, "Grid resource discovery based on distributed learning automata," *Computing*, vol. 96, no. 9, pp. 909–922, 2014.

54 C. Di, Q. Liang, F. Li, S. Li, and F. Luo, "An efficient parameter-free learning automaton scheme," *IEEE Transactions on Neural Networks and Learning Systems*, vol. 32, no. 11, pp. 4849–4863, 2020.

55 H. Lu, "Function optimization-based schemes for designing continuous action learning automata," Ph.D. dissertation, Université d'Ottawa/University of Ottawa, 2019.

56 W. Liu, L. Dong, D. Niu, and C. Sun, "Efficient exploration for multi-agent reinforcement learning via transferable successor features," *IEEE/CAA Journal of Automatica Sinica*, vol. 9, no. 9, pp. 1673–1686, 2022.

57 J. R. Wang, Y. T. Hong, J. L. Wang, J. P. Xu, Y. Tang, Q.-L. Han, and J. Kurths, "Cooperative and competitive multi-agent systems: From optimization to games," *IEEE/CAA Journal of Automatica Sinica*, vol. 9, no. 5, pp. 763–783, 2022.

58 L. Xia, Q. Li, R. Song, and H. Modares, "Optimal synchronization control of heterogeneous asymmetric input-constrained unknown nonlinear mass via reinforcement learning," *IEEE/CAA Journal of Automatica Sinica*, vol. 9, no. 3, pp. 520–532, 2022.

59 L. Huang, M. Zhou, K. R. Hao, and E. Hou, "A survey of multi-robot regular and adversarial patrolling," *IEEE/CAA Journal of Automatica Sinica*, vol. 6, no. 4, pp. 865–874, 2019.

60 L. Huang, M. Zhou, and K. Hao, "Non-dominated immune-endocrine short feedback algorithm for multi-robot maritime patrolling," *IEEE Transactions on Intelligent Transportation Systems*, vol. 21, no. 1, pp. 362–373, 2020.

61 L. Huang, M. Zhou, K. Hao, and H. Han, "Multirobot cooperative patrolling strategy for moving objects," *IEEE Transactions on Systems, Man, and Cybernetics: Systems*, vol. 53, no. 5, pp. 2995–3007, 2023.

62 Z. Zhang, D. Wang, and J. Gao, "Learning automata-based multiagent reinforcement learning for optimization of cooperative tasks," *IEEE Transactions on Neural Networks and Learning Systems*, vol. 32, no. 10, pp. 4639–4652, 2020.

63 Z. Zhang, H. Liu, M. Zhou, and J. Wang, "Solving dynamic traveling salesman problems with deep reinforcement learning," *IEEE Transactions on Neural Networks and Learning Systems*, vol. 34, no. 4, pp. 2119–2132, 2023.

64 C. Lin, Z. Cao, and M. Zhou, "Learning-based grey wolf optimizer for stochastic flexible job shop scheduling," *IEEE Transactions on Automation Science and Engineering*, vol. 19, no. 4, pp. 3659–3671, 2022.

65 C. Lin, Z. Cao, and M. Zhou, "Learning-based cuckoo search algorithm to schedule a flexible job shop with sequencing flexibility," *IEEE Transactions on Cybernetics*, vol. 53, no. 10, pp. 6663–6675, 2023.

Index

Learning Automata and Their Applications to Intelligent Systems, First Edition.
JunQi Zhang and MengChu Zhou.
© 2024 The Institute of Electrical and Electronics Engineers, Inc. Published 2024 by John Wiley & Sons, Inc.

Printed and bound by CPI Group (UK) Ltd, Croydon, CR0 4YY

16/04/2025

14658581-0001